T0185574

SpringerBriefs in Crystallography

Editor-in-Chief

Massimo Nespolo, CRNS, CRM2, Université de Lorraine, Nancy, France

Series Editors

Livio Battezzati, University of Turin, Dipartimento di Chimica, Turin, Italy

Gautam Desiraju, Indian Institute of Science Bangalore, Bangalore, Karnataka, India

Mike Glazer, Claredon Lab, Department of Physics, University of Oxford, Oxford, UK

Joke Hadermann, Department of Physics, Universiteit Antwerpen, Antwerpen, Belgium

Dorothee Liebschner, Lawrence Berkeley National Laboratory, Berkeley, CA, USA

Koichi Momma, National Museum of Nature and Science, Tsukuba-shi, Ibaraki, Japan

Berthold Stöger, X-ray center, Technische Universität Wien, Vienna, Austria

SpringerBriefs in Crystallography, published under the auspices of the International Union of Crystallography, aims at presenting highly relevant, concise monographs with an intermediate scope between a topical review and a full monograph. Areas of interest include chemical crystallography, crystal engineering, crystallography of materials (ceramics, metals, organometallics, functional materials), instrumentation, mathematical crystallography, mineralogical crystallography, physical properties of crystals, structural biology and related fields.

SpringerBriefs present succinct summaries of cutting-edge research and practical applications covering a range of content from professional to academic and featuring compact volumes of 50 to 125 pages.

- A timely report of state-of-the art experimental techniques and instrumentation
- New computation algorithms or theoretical approaches
- A bridge between new research results, as published in journal articles, and a contextual literature review
- A snapshot of a hot or emerging topic
- An in-depth case study
- A presentation of core concepts that students must understand in order to make independent contributions

Briefs are characterized by fast, global electronic dissemination, standard publishing contracts, standardized manuscript preparation and formatting guidelines, and expedited production schedules.

Publications in these series help support the Outreach and Education Fund for the International Union of Crystallography.

More information about this series at http://www.springer.com/series/16236

Luca Bindi

Natural Quasicrystals

The Solar System's Hidden Secrets

 Springer

Luca Bindi
Dipartimento di Scienze della Terra
Università degli Studi di Firenze
Florence, Italy

Sezione di Firenze
Istituto di Geoscienze e Georisorse, C.N.R.
Florence, Italy

ISSN 2524-8596   ISSN 2524-860X   (electronic)
SpringerBriefs in Crystallography
ISBN 978-3-030-45676-4   ISBN 978-3-030-45677-1   (eBook)
https://doi.org/10.1007/978-3-030-45677-1

© The Author(s), under exclusive license to Springer Nature Switzerland AG 2020
This work is subject to copyright. All rights are solely and exclusively licensed by the Publisher, whether the whole or part of the material is concerned, specifically the rights of translation, reprinting, reuse of illustrations, recitation, broadcasting, reproduction on microfilms or in any other physical way, and transmission or information storage and retrieval, electronic adaptation, computer software, or by similar or dissimilar methodology now known or hereafter developed.
The use of general descriptive names, registered names, trademarks, service marks, etc. in this publication does not imply, even in the absence of a specific statement, that such names are exempt from the relevant protective laws and regulations and therefore free for general use.
The publisher, the authors and the editors are safe to assume that the advice and information in this book are believed to be true and accurate at the date of publication. Neither the publisher nor the authors or the editors give a warranty, expressed or implied, with respect to the material contained herein or for any errors or omissions that may have been made. The publisher remains neutral with regard to jurisdictional claims in published maps and institutional affiliations.

This Springer imprint is published by the registered company Springer Nature Switzerland AG
The registered company address is: Gewerbestrasse 11, 6330 Cham, Switzerland

Dedicated to my mother.

Nothing in this world was so strong and enduring as her love.

Acknowledgements

First of all, I wish to thank my family, friends, and scientific collaborators for sharing their love, support, and brilliance with me. This short monograph reviews work that has benefitted greatly from my valued collaborators: C. L. Andronicos, P. D. Asimow, H. Buscmann, J. Eiler, V. Distler, M. P. Eddy, Y. Fei, Y. Guan, P. R. Heck, R. Hemley, L. S. Hollister, J. Hu, A. Kostin, V. Kryachko, C. Lin, P. Lu, C. Ma, H.-K. Mao, M. M. M. Meier, G. J. MacPherson, J. Oppenheim, J. Pham, G. R. Poirier, W. Steinhardt, V. Stagno, E. Stolper, O. Tschauner, N. Yao, R. Wieler, and M. Yudovskaya.

My greatest, warmest thanks, however, goes to my friend Paul J. Steinhardt, the person who lived with me day by day every moment of the "natural quasicrystals" saga and the person with whom I began daily communication in 2008 and have been brainstorming and planning together each day since.

Last but not least, I wish to thank Massimo Nespolo, for the invitation to write the first SpringerBrief of this series, for his line-by-line careful review, and for his critical approach to crystallographic contributions in general and to mineralogical structural complexities in particular. Critical reviews by Livio Battezzati, Koichi Momma, and John A. Jaszczak helped to improve the clarity of the text.

The author wishes to thank MIUR-PRIN2017, project "TEOREM deciphering geological processes using Terrestrial and Extraterrestrial ORE Minerals", prot. 2017AK8C32.

Contents

Chapter 1
Introduction

The theoretical concept of quasicrystals (Levine and Steinhardt 1984) and the contemporary experimental breakthrough (Shechtman et al. 1984) happened about 35 years ago. Since then, several new types of quasicrystalline materials have been synthesized in the laboratory under controlled protocols designed to avoid solidification of material having a periodic structure typical of a crystal (Steurer and Deloudi 2009; Steurer 2018). Levine and Steinhardt (1984) pointed out that quasicrystals can theoretically be as robust and stable as crystals, perhaps even forming in Nature. On these grounds, a more-than-decade-long search for a natural quasicrystal began, which ended with the discovery of *icosahedrite* ($Al_{63}Cu_{24}Fe_{13}$), an icosahedral quasicrystal found in a specimen from the Koryak Mountains, Far Eastern Russia.

In this brief monograph the author wants to highlight the search and discovery, the laboratory experiments/analyses showing the sample to be extraterrestrial, the results of an extraordinary geological expedition to the Koryak Mountains to pursue further evidence, and the experimental proof supporting the formation of such exotic materials in hypervelocity collisions among extraterrestrial objects in outer space million years ago. The occurrence of natural quasicrystals inside a meteorite (together with several stable crystalline phases) demonstrates that quasicrystals can form naturally within a composite, inhomogeneous medium. The result not only settles the fundamental issues about quasicrystals as being a stable form of matter, but also shows how quasicrystals formed at the beginning of the Solar System and, though once thought to be impossible, are probably not such rare materials in the Milky Way, and perhaps elsewhere in the Universe. Indeed, natural periodic approximants, crystalline compounds having structures obeying the classic laws of 3D-crystallography, have been already described—even if the authors did not notice it—from Moon (the mineral naquite, FeSi; Anand et al. 2004), cometary interplanetary dust particles (the mineral brownleeite, MnSi; Nakamura-Messenger et al. 2010) and diamond inclusions (unnamed Mn–Ni–Si-alloys; Galimov et al. 2020).

© The Author(s), under exclusive license to Springer Nature Switzerland AG 2020
L. Bindi, *Natural Quasicrystals*, SpringerBriefs in Crystallography,
https://doi.org/10.1007/978-3-030-45677-1_1

The composition of natural quasicrystals possibly represents the most important *Solar System secret* hidden in these compounds. Indeed, the presence of metallic Al requires enormously high reducing conditions to form, even more stringent than those hypothesized for the innermost hot regions of the pre-solar nebula 4.567 Gy ago (Grossman et al. 2008). Moreover, Al and Cu (both coexisting in the first discovered natural quasicrystal; Bindi et al. 2009) show a totally different cosmochemical behavior: Al is a lithophile element, which condenses from a hot and cooling gas of Solar composition at very high temperatures; on the contrary, Cu is a siderophile/chalcophile moderately-volatile element condensing at much lower temperatures. How did they come together? This represents one of the major conundrums for the scientific community to be deciphered in the time to come.

The discovery of quasicrystals in nature (Bindi et al. 2009) and the fact that several authors described potential quasicrystal candidates (or quasicrystal approximants) without realizing their nature, showed us that minerals exhibiting such complex structures could be more common than thought. They could also be the tip of a huge iceberg of novel materials to be discovered in the mineral kingdom. Our capacity to discover them in Nature could be limited only by psychological barriers and by people who do not disregard 'crystallographic oddities' a priori. My hope is to see in the near future many new classes of novel materials that have yet to be envisioned. Mineral sciences in general, and mineralogical crystallography in particular, can continue to amaze us and have a strong influence on other disciplines, including chemistry, material science, solid state physics, engineering, cosmo-chemistry and astrophysics. Minerals could indeed be the treasure trove of other *secrets* of our (and other) *Solar System(s)*.

References

Anand M, Taylor LA, Nazarov MA, Shu J, Mao H-K, Hemley RJ (2004) Space weathering on airless planetary bodies: clues from the lunar mineral hapkeite. Proc Natl Acad Sci USA 101:6847–6851

Bindi L, Steinhardt PJ, Yao N, Lu PJ (2009) Natural quasicrystals. Science 324:1306–1309

Galimov EM, Kaminsky FV, Shilobreeva SN, Sevastyanov VS, Voropaev SA, Khachatryan GK, Wirth R, Schreiber A, Saraykin VV, Karpov GA, Anikin LP (2020) Enigmatic diamonds from the Tolbachik volcano, Kamchatka. Am Mineral 105:498–509

Grossman L, Beckett JR, Fedkin AV, Simon SB, Ciesla FJ (2008) Redox conditions in the solar nebula: observational, experimental, and theoretical constraints. Rev Mineral Geochem 68:93–140

Levine D, Steinhardt PJ (1984) Quasicrystals: a new class of ordered structures. Phys Rev Lett 53:2477–2480

Nakamura-Messenger K, Keller L, Clemett SJ, Messenger S, Jones JH, Palma RL, Pepin RO, Klock W, Zolensky ME, Tatsuoka H (2010) Brownleeite: a new manganese silicide mineral in an interplanetary dust particle. Am Mineral 95:221–228

Shechtman D, Blech I, Gratias D, Cahn JW (1984) Metallic phase with long-range orientational order and no translational symmetry. Phys Rev Lett 53:1951–1953

Steurer W (2018) Quasicrystals: what do we know? What do we want to know? What can we know? Acta Crystallogr A 74:1–11

Steurer W, Deloudi S (2009) Crystallography of quasicrystals. In: Concepts, methods and structures. Springer series in materials science, vol 126. Springer, Heidelberg

Chapter 2
What Are Quasicrystals and Why They Are so Important?

The fact that chemical elements combine to form crystals, periodic objects where the atoms are arranged in a periodic lattice of points with a limited set of symmetries, has been a basic belief for more than two centuries. Through the time, the rigorous mathematical theorems of crystallography have been codified in a set of principles known as *the laws of crystallography*, which govern the outcome we expect when measuring the physical properties of materials, whether it be making silver for a bracelet, cutting the facets of a precious gem or working copper to prepare electrical wires. According to the classic understanding, atoms are either randomly distributed, as in the case of gases, or crystalline, as is the case for salt or quartz. In crystalline materials, the atoms form symmetrical structures like hexagons in a tiling, which repeat periodically exhibiting a distinct rotational symmetry. Periodic and non-periodic materials are comparable to mosaics, where the tiles can be assembled together randomly or in an orderly, symmetrical tessellation. A crucial rule for the regular tessellations, known since the ancient Egyptians, is that you can use only determinate shapes for the tiles and, even more important, not all the symmetries can be obtained. The analogy can be transposed to solid matter. Thus, periodic materials can be only governed by certain rotational symmetries: one-, two-, three-, four- and six-fold symmetry axes in two and three dimensions; five-, seven-, eight- or higher-fold symmetry axes are strictly forbidden.

In 1984, Shechtman et al. (1984) published the exceptional discovery of a piece of synthetic material with the Al_6Mn composition having a diffraction pattern of sharp spots like a crystal, but with the symmetry of an icosahedron. The icosahedral symmetry is forbidden because it includes six independent five-fold symmetry axes. As it happens for the most important breakthroughs in science, a theoretical explanation for this astonishing experimental evidence appeared the same year. Indeed, in the same months of 1984, Dov Levine and Paul J. Steinhardt at the University of Pennsylvania had been developing the idea of a new form of solid they dubbed *quasicrystals*, short for quasiperiodic crystals, where a *quasiperiodic* atomic arrangement means that the atomic positions can be described by a sum of periodic functions whose periods have an irrational ratio (Levine and Steinhardt 1984).

© The Author(s), under exclusive license to Springer Nature Switzerland AG 2020
L. Bindi, *Natural Quasicrystals*, SpringerBriefs in Crystallography,
https://doi.org/10.1007/978-3-030-45677-1_2

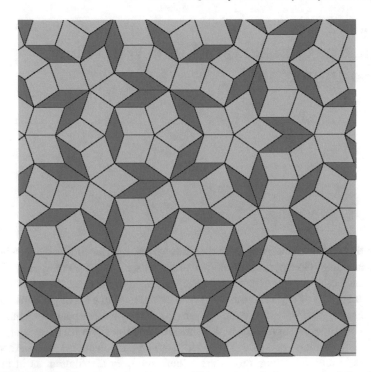

Fig. 2.1 Portion of a two-dimensional Penrose tiling composed of two types of tiles (blue and green) arranged with a crystallographically-forbidden five-fold symmetry

Quasiperiodic materials[1] can show strong and sharp diffraction peaks as the best of the crystalline materials but can violate the mathematical theorems that constrain crystal symmetries. A classic 2D-example is the Penrose tiling (Penrose 1974), which shows how two tiles (flat and slim rhombi) are arranged in a five-fold symmetric pattern (Fig. 2.1).

Quasicrystals can indeed exhibit several forbidden symmetries, including icosahedral symmetry. Levine and Steinhardt (1984) demonstrated that the tiles in the Penrose tiling actually repeat with incommensurate frequencies each to other; analogously, the same *incommensurability* could be used to construct polyhedral units with protrusions and holes on their faces that constrain the way they join together such that the units can only fit together in a three-dimensional solid with icosahedral symmetry (Fig. 2.2).

Noteworthy, the diffraction pattern hypothesized by Levine and Steinhardt perfectly matched that obtained experimentally by Shechtman et al. (1984). This model

---

[1]In this short monograph, I will simply report the basics to understand what a quasicrystal is. For a more comprehensive review, including the description of quasicrystalline structures in the superspace using the higher-dimensional crystallography, the readers can refer to general reviews as those given by Steurer and Deloudi (2009), Janssen et al. (2007), van Smaalen (2007) and Bindi and Chapuis (2017).

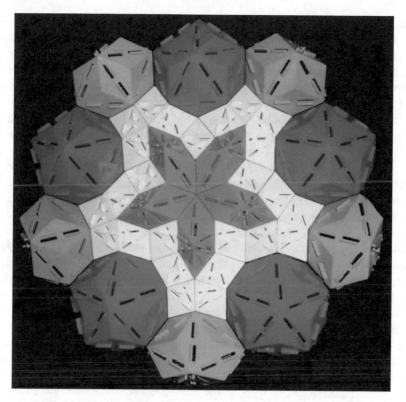

Fig. 2.2 Portion of a 3D icosahedral quasicrystal made of four types of polyhedral units and built in a way that assure to all space-filling arrangements to be quasicrystalline. For details see Socolar and Steinhardt (1986). Modified after Steinhardt and Bindi (2012)

has been criticized, though, because the rules seemed too complicated to be manifested in the simple atomic interactions of metal alloys. Other people questioned either the experimental evidence by Shechtman et al. (1984) or the theoretical explanation given by Levine and Steinhardt (1984). Among them, the double Nobel laureate Linus Pauling did not accept at the beginning the quasiperiodicity as an actual state: '*Apparent icosahedral symmetry is due to directed multiple twinning of cubic crystals*' (Pauling 1985). However, as it has been well documented by Steurer and Deloudi (2009), the increasing quality of the synthesized quasicrystalline materials together with an ever better quality of the collected X-ray data, forced Pauling to propose increasingly laborious hypotheses to validate his twinning models [from initial models using 1120 atoms (Pauling 1985) up to 19,400 atoms per unit cell (Pauling 1989)]. Pauling and his apologists always rejected the idea to transform the three dimensional non-crystallographic five-fold symmetry into a higher-dimensional crystallographic one—which involves, for a practical description, to consider the quasicrystalline structure as a projection in the 3-dimensional physical space (Neubüser et al. 1971). Pauling's and other alternative proposals lost out in

1988, however, mainly because of the discovery of a different alloy that was unquestionably a perfect quasicrystal (Tsai et al. 1987), and which can be considered the first *bona fide* quasicrystal. Since then, more than one hundred quasicrystalline materials have been synthesized, usually starting from specific proportions of the elements at relatively high temperature, and then quenching following precise protocols (Janot 1994).

Finally, after years of doubts and scepticism, the 2011 Nobel Prize in Chemistry was awarded to Dan Shechtman for his experimental breakthrough that changed our thinking about possible forms of solid matter.

Noteworthy, today the definition of quasicrystals as materials possessing an axis of symmetry that is incompatible with periodicity is obsolete. Indeed, according to this restriction there should not be quasicrystals in 1-dimension, and a quasicrystal in 2- or 3-dimensions should have an axis of N-fold symmetry, with $N = 5$, or $N > 6$. Lifshitz (2003) proposed adopting the original definition of Levine and Steinhardt (1984) whereby the term *quasicrystal* is simply an abbreviation for *quasiperiodic crystal*, possibly with the condition that the term quasicrystal be used for crystals that are strictly aperiodic (as the mathematical definition of quasiperiodicity includes periodicity as a special case).

The quasiperiodic translational order of quasicrystals has, obviously, physical consequences. Electrons and phonons in these materials do not come across to a periodic potential, and so they can exhibit infrequent resistive and elastic properties (Dubois 2005). As reviewed by Lu (2000), single metallic elements (typically highly conducting) have a completely different behaviour when are alloyed to form a quasicrystal (characterized by a very high electrical resistance; Rapp 1998). Furthermore, when a single, periodic metallic element is heated at high temperature, a disorder is observed in the crystal structure accompanied by an increase in the electrical resistivity. By contrast, in a quasicrystal, the disorder induced by heating dramatically decreases its electrical resistivity (Rapp 1998). Unlike the soft crystalline metals, quasicrystal alloys are exceedingly hard and have a high degree of surface slipperiness, motivating the first commercial application of quasicrystals as a cookware coating alternative to Teflon (Gibbons and Kelton 1998). The forbidden symmetry of 12-fold symmetric quasicrystals has been exploited to create novel photonic waveguides (Zoorob et al. 2000) and, more recently, in the realization of photonic quasicrystals that can influence optical transmission and reflectivity, photoluminescence, light transport, plasmonics and laser action (Valy Vardeny et al. 2013 and references therein).

References

Bindi L, Chapuis G (2017) Aperiodic mineral structures. In: Plasil J, Majzlan J, Krivovichev S (eds) Mineralogical crystallography. EMU notes in mineral, vol 19, pp 213–254
Dubois J-M (2005) Useful quasicrystals. World Scientific, 482 pp

Gibbons PC, Kelton KF (1998) Toward industrial applications. In: Stadnik ZM (ed) Physical properties of quasicrystals. Springer-Verlag

Janot C (1994) Quasicrystals: a primer. Oxford University Press, Oxford

Janssen T, Chapuis G, de Boissieu M (2007) Aperiodic crystals: from modulated phases to quasicrystals. Oxford University Press, Oxford

Levine D, Steinhardt PJ (1984) Quasicrystals: a new class of ordered structures. Phys Rev Lett 53:2477–2480

Lifshitz R (2003) Quasicrystals: a matter of definition. Found Phys 33:1703–1711

Lu PJ (2000) The search for new quasicrystals. Bachelor thesis in physics, Princeton University, 160 pp

Neubüser J, Wondratschek H, Bülow R (1971) On crystallography in higher dimensions. I. General definitions. Acta Crystallogr A 27:517–520

Pauling L (1985) Apparent icosahedral symmetry is due to directed multiple twinning of cubic crystals. Nature 317:512–514

Pauling L (1989) Icosahedral quasicrystals of intermetallic compounds are icosahedral twins of cubic crystals of three kinds, consisting of large (about 5000 atoms) icosahedral complexes in either a cubic body-centered or a cubic face-centered arrangement or smaller (about 1350 atoms) icosahedral complexes in the b-tungsten arrangement. Proc Natl Acad Sci USA 86:8595–8599

Penrose R (1974) The role of aesthetics in pure and applied mathematical research. Bull Inst Math Appl 10:266–271

Rapp O (1998) Electronic transport properties of quasicrystals—experimental results. In: Stadnik ZM (ed) Physical properties of quasicrystals. Springer-Verlag

Shechtman D, Blech I, Gratias D, Cahn JW (1984) Metallic phase with long-range orientational order and no translational symmetry. Phys Rev Lett 53:1951–1953

Socolar J, Steinhardt PJ (1986) Quasicrystals II: unit cell configurations. Phys Rev B 34:617–647

Steinhardt PJ, Bindi L (2012) In search of natural quasicrystals. Rep Prog Phys 75:092601–092611

Steurer W, Deloudi S (2009) Crystallography of quasicrystals: concepts, methods and structures. Springer series in materials science, vol 126. Springer, Heidelberg

Tsai AP, Inoue A, Masumoto T (1987) A stable quasicrystal in Al-Cu-Fe system. Jpn J Appl Phys 26:L1505

Valy Vardeny M, Nahata A, Agrawal A (2013) Optics of photonic quasicrystals. Nat Photonics 7:177–187

van Smaalen S (2007) Incommensurate crystallography. Oxford University Press, Oxford

Zoorob ME, Charlton MDB, Parker GJ, Baumberg JJ, Netti MC (2000) Complete photonic bandgaps in 12-fold symmetric quasicrystals. Nature 404:740–743

Chapter 3
Can Nature Have Beaten Us to the Punch?

After the discovery of an unquestionably perfect quasicrystal by Tsai et al. (1987), a new debate immediately began: Why do quasicrystals form? Are they actually stable forms of matter like crystals? Some have argued that all quasicrystals are inherently delicate, metastable oddities that must be synthesized under highly controlled laboratory conditions (Henley 1991). By contrast, Levine and Steinhardt (1984) showed that, in principle, quasicrystals could be as stable and robust as crystals. According to the latter hypothesis, is it possible that Nature had beat us to the punch by forming quasicrystals long before they were made in the laboratory? There were numerous motivations for pursuing this question. For geoscience, the discovery of a natural quasicrystal would have opened a new chapter in the study of mineralogy, forever altering the conventional classification of mineral forms. For solid state physics, to find a natural quasicrystal would have pushed back the age of the oldest quasicrystal (the Shechtman's material) by orders of magnitude and would have helped to elucidate either how these materials form or their stability over extraordinary long annealing times. Identifying new compositions able to give quasicrystalline materials has always relied significantly on serendipity, and searching through Nature could prove to be an effective complement to laboratory methods. Finally, and maybe the most fascinating aspect, the eventual discovery of quasicrystals in Nature could suggest new geologic or extra-terrestrial processes.

3.1 Search and Discovery

Several informal searches through the mineralogical collections of major museum began soon after the first synthetic quasicrystal was discovered in the laboratory (Shechtman et al. 1984), but they yielded no results. Then, nearly two decades ago, a systematic search (Lu et al. 2001) was initiated using a novel scheme for identifying quasicrystals based on powder diffraction data. With this search Lu et al. (2001)

© The Author(s), under exclusive license to Springer Nature Switzerland AG 2020
L. Bindi, *Natural Quasicrystals*, SpringerBriefs in Crystallography,
https://doi.org/10.1007/978-3-030-45677-1_3

identified about fifty promising candidates (among them there were 6 minerals) that were further studied by TEM and X-ray diffraction, but, in the end, no new quasicrystals, synthetic or natural, were discovered in the original study. At that point, a world-wide search through museums and private collections eventually led to the discovery of a promising candidate belonging to the mineralogical collections of the Museo di Storia Naturale (Natural History Museum) of the University of Florence, thanks to the intuition to test minerals whose compositions were similar to known quasicrystals synthesized in the laboratory. The sample, labelled "khatyrkite" (catalogue number 46407/G; Fig. 3.1), was acquired by the Florence museum in 1990 from a private collector and catalogued as coming from the Khatyrka region of the Koryak mountains in the Chukotka autonomous Okrug on the north-eastern part of the Kamchatka peninsula (Bindi et al. 2009, 2011, 2012).

Khatyrkite, ideally $(Cu,Zn)Al_2$, and cupalite, ideally $(Cu,Zn)Al$, are two minerals first described by Razin et al. (1985). In the 46407/G sample (hereafter the "Florence sample"), khatyrkite is intergrown with typical rock-forming minerals (e.g., forsterite and diopside), other metallic crystalline phases (cupalite and an unnamed β-AlCuFe phase), and a few grains of a new species, with composition $Al_{63}Cu_{24}Fe_{13}$,

4 mm

Fig. 3.1 Views in different orientations of the khatyrkite sample belonging to the collections of the Museo di Storia Naturale of the Università degli Studi di Firenze (catalogue number 46407/G). The lighter-coloured material on the exterior contains a mixture of spinel, clinopyroxene and olivine. The dark material consists predominantly of khatyrkite (ideally $CuAl_2$) and cupalite (ideally $CuAl$), but also includes granules of icosahedrite with composition $Al_{63}Cu_{24}Fe_{13}$. Modified after Bindi and Steinhardt (2014)

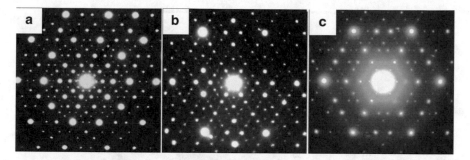

Fig. 3.2 The unambiguous signature of an icosahedral quasicrystal consists of patterns of sharp peaks arranged in straight lines in an incommensurate structure with five-fold (**a**), two-fold (**b**) and three-fold (**c**) symmetry. The electron diffraction patterns shown here, taken from a grain of icosahedrite, match those predicted for a face-centered icosahedral quasicrystal, as do the angles that separate the symmetry axes. Modified after Bindi et al. (2009)

whose X-ray powder diffraction pattern did not match that of any known mineral. The $Al_{63}Cu_{24}Fe_{13}$ unnamed species was found to have a perfect icosahedral quasicrystalline structure (Fig. 3.2).

3.2 Degree of Structural Perfection

The electron diffraction patterns obtained with a transmission electron microscope showed sharp peaks arranged in an incommensurate structure with five-, three- and two-fold symmetry (Fig. 3.2), the characteristic signature of an icosahedral quasicrystal (Janot 1994). In addition, the angles between the symmetry planes are consistent with icosahedral symmetry. For example, the angle between the two- and five-fold symmetry planes was measured to be 31.6(5)°, which agrees with the ideal angle between the two-fold and five-fold axes of an icosahedron ($\tan^{-1} 1/\tau \approx 31.7°$).

The electron diffraction study also demonstrated the high degree of structural perfection of the natural quasicrystal. Quasicrystals produced in the laboratory by rapid quenching and/or embedded in a matrix of another phase often exhibit measurable phason strains (Levine et al. 1985; Lubensky et al. 1986), where with the term "phason" we refer to a quasiparticle existing in quasicrystals, and corresponds to a specific type of disorder relative to an ideal quasiperiodic structure. Similar to a phonon, a phason is associated with atomic motion. However, whereas phonons are related to *translation* of atoms, phasons are associated with atomic *rearrangements*. In the Penrose tiling of Fig. 3.3, for example, a phason would result in flipping of local groups of tiles. As a result of these rearrangements, waves describing the position of atoms in crystals change phase, thus the term "phason". If the density $\rho(x)$ is decomposed into a sum of incommensurate density waves, a phason strain is a gradient in the phason shift between incommensurate density waves (Lubensky et al. 1985). An experimental signature of a phason is a shift in the Bragg-peak

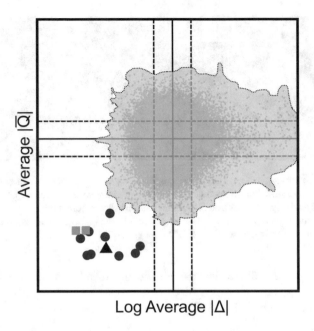

Fig. 3.3 Plot showing the distribution of two figures-of-merit given by Lu et al. (2001) to differentiate quasicrystals from the large collection of powder patterns in the ICDD-PDF: the logarithm of the intensity-weighted average $|\Delta|$, where Δ is the absolute deviation of each Q from the closest-matching face-centered-icosahedral peak and the intensity-weighted average of $\bar{Q}$. Known synthetic face-centered-icosahedral quasicrystals are indicated with grey squares (i-AlCuFe) and dark grey circles (other examples). They cluster far from ordinary crystalline minerals (grey area), whose average and standard deviation are indicated by solid and dashed lines, respectively. The natural quasicrystal, marked with the black triangle, is several standard deviations away from the average and well within the cluster of known face-centered-icosahedral quasicrystals. Modified after Bindi et al. (2009)

position from its ideal position by an amount proportional to $\bar{Q}$, corresponding to larger shifts for peaks with the smaller intensity.[1] By holding a diffraction pattern at a grazing angle and viewing down rows of peaks, the phason strain can be observed as deviations of the weaker peak positions from straight rows (Lubensky et al. 1986). The electron diffraction patterns in Fig. 3.2 display no discernible evidence of phason strain. Based on past laboratory experience, forming quasicrystals with such a high degree of perfection under these complex conditions seems nearly impossible. Similarly, the score on the $\bar{Q}$ figure-of-merit test (Lu et al. 2001) that is a sensitive test of phason strain, calculated using the powder diffraction pattern obtained for the

---

[1] The diffraction pattern of an ideal three-dimensional quasicrystal consists of Bragg peaks located on a lattice given by $\bar{Q} = \sum_{i=1}^{6} n_i b_i$ where the b_i are basis vectors pointing to the vertices along the six five-fold symmetry axes of a regular icosahedron in three dimensions, and the n_i are integers that index each vector. The b_i vectors were chosen as follows: $b_1 = (1, \tau, 0)$, $b_2 = (\tau, 0, 1)$, $b_3 = (0, 1, \tau)$, $b_4 = (-1, \tau, 0)$, $b_5 = (\tau, 0, -1)$, $b_6 = (0, -1, \tau)$, where τ is the golden ratio, $(1 + \sqrt{5})/2$.

natural quasicrystal (Bindi et al. 2009), demonstrates that the mineral has a degree of structural perfection comparable to the best laboratory specimens (Fig. 3.3).

Either the quasicrystalline mineral samples formed without phason strain in the first place, or subsequent annealing was sufficient for phason strains to relax away.

3.3 Does the First Natural Quasicrystal Deserve a Name?

Historically, all known natural minerals with translational order have been crystals or incommensurate crystals with rotational symmetries restricted to the finite set of crystallographic possibilities established mathematically in the nineteenth century. In this light, the first natural quasicrystal clearly presented the first exception: a three-dimensional icosahedral symmetry strictly forbidden for crystals. Moreover, it also showed a new chemistry, $Al_{63}Cu_{24}Fe_{13}$, never observed for other natural substances. According to the rules of the International Mineralogical Association, when a mineral with a new chemistry and a new structure is found in nature, it deserves a new name (to be included in the list of substances formed by Nature). We named the new mineral *icosahedrite* for the icosahedral symmetry of its atomic structure, which is clearly evident from diffraction data (Fig. 3.2). The new mineral and mineral name have been approved by the Commission on New Minerals, Nomenclature and Classification of the International Mineralogical Association (Bindi et al. 2011).

3.4 How Had Nature Done It? Impossible...

After the finding of the first quasicrystal in Nature and the successful proposal of icosahedrite as new mineral to the International Mineralogical Association, the story might have ended there, if not for an enigma. The conditions to form the mineral were nothing like the pristine conditions used to synthesize quasicrystals in the laboratory. How could Nature have done it? Paul Steinhardt, a Princeton physicist and my valued partner in the search for the first quasicrystal in Nature, approached the famous petrologist Lincoln S. Hollister (Princeton University) to find out how this might be possible, the response was short and to the point: *it's impossible!* The quasicrystal contained metallic aluminum, Hollister noted, which can only be separated from oxygen artificially, as in aluminum processing plants. The sample had to be slag, a worthless man-made artifact. The apparent success had now turned to depressing failure, leading to the next chapter in the story.

3.5   Farfetched Coincidences

For more than a year-and-a-half, we worked to trace the origin of the Florence sample. During this period, we had the impression of living a fairytale; yet, every bit of the story is incredibly true. As in any good fairytale, impossible things happen and impossible barriers are overcome again and again, sometimes magically by good luck and sometimes through fanatical determination. The documentation in the museum related to the Florence sample reported that it made part of a large micromounts collection bought from Nicholas Koekkoek, a private mineral collector living in Amsterdam with no presence on the Internet and a common Dutch family name. After several detective searches, we did not find any trace of Koekkoek, but something impossible was going to happen. I was recounting the story of the discovery of quasicrystals in nature to some friends during a dinner nine years ago when one of them (who lives in Netherlands) recalled that an old woman named Koekkoek lived close to him in Amsterdam. When my Dutch friend returned home, he asked her neighbor if she knew a mineral collector who shared her surname. Bizarrely, Nicholas Koekkoek was the old woman's deceased husband! I immediately reserved a flight to Amsterdam to meet the widow. When I met her, she (of course) did not know anything about the Florence sample—she did not share the hobby of minerals with his husband. However, she was so kind to show me her husband's *secret diary* where the collector diligently annotated all the samples he had purchased/exchanged during his life. At the page containing information on the Florence sample (it was easy to locate it because I knew the catalogue number originally used by Koekkoek), Koekkoek explained how he purchased the mineral sample. It was bought in 1987 from a man named Tim during a trip to Romania. Immediately, the question was: where had Tim obtained the sample? I spent more than two months trying to find him, but nothing. Thus, with the hope to find out more 'hidden' information in the *secret diary*, I decided to take another flight to Amsterdam to meet again the Koekkoek's widow. After several discussions (she did not like very much to discuss of her husband), and thanks to my incredible shameless stubbornness, she gave up and confessed to me the existence of a *secret diary*. There, I learnt that Koekkoek had actually obtained the small sample from Leonid Razin, labeled in the diary as the Director of the Institute of Platinum in St. Petersburg, Russia. I recognized that name. Razin was actually the first author of the 1985 original publication describing and characterizing khatyrkite and cupalite as new minerals (Razin et al. 1985). Razin and his coauthors then deposited the holotype in the collections of the Mining Institute in St. Petersburg. Although we are not certain, it seems that the holotype and the Florence sample were collected together in the field, and that the first was studied by Razin et al. and the latter was sold. Paul Steinhardt was able to talk (at the telephone) with Leonid Razin to ask for more information but, unfortunately, he could not remember all the details that happened almost 30 years earlier. We were stuck again. Having no precise ideas, I started wandering though the scientific literature, specifically looking at the 1985 paper by Razin and coauthors. With my surprise, I noticed that the first paragraph revealed a name, Valery Kryachko, who seemed to have played a key role in the finding of the holotype sample. I started to write letters to all my Russian collaborators and learnt that Kryachko was probably an untraceable rural miner who likely found the sample then studied by Razin and coauthors while looking for platinum and gold nuggets. But the surprises were not over. By chance, I found the Kryachko's name among the authors of other papers published in the period 1995–1999. To make a very long story very short, we were able to locate Kryachko. He explained that he actually collaborated with Razin and that in the late seventies he was brought by helicopter at the Listvenitovyi stream where he spent a period to look for platinum digging through the blue-green clays. No platinum was found, only a few shiny small nuggets he was not able to identify. He delivered them to Razin and never heard about them in the subsequent years.

3.6 An Unexpected Origin: They Come from Outer Space

One of the *game changer* to decipher the origin of the Florence sample was the discovery of a very tiny fragment (50 µm in size; Fig. 3.4) of the mineral stishovite, the high-pressure polymorph of SiO_2 (Bindi et al. 2012). It crystallizes in the space-group type $P4_2/mnm$ and has unit-cell parameters $a \approx 4.2$ Å and $c \approx 2.7$ Å.

Differently from what is commonly observed in silicates having Si in tetrahedral coordination, in the mineral stishovite Si is octahedrally coordinated to six oxygen atoms, symptomatic of a very high-pressure regime of formation. Indeed, it is usually produced at shock pressures ≥ 10 GPa and temperatures ≥ 1250 °C. The presence of stishovite in the Florence sample strongly indicated a formation in a very peculiar environment: Earth's lower mantle or in a collision in space among asteroids. Remarkably, the stishovite contains inclusions of icosahedrite, an indication that it also formed at high pressure.

Fig. 3.4 TEM image of an icosahedrite inclusion in stishovite

2 nm

Stishovite is commonly found in shock-metamorphosed meteorites, and so we thought that the Florence sample could be extraterrestrial in origin. But how to prove it? Fortunately, it is possible to distinguish between terrestrial and extraterrestrial materials. Indeed, stable isotopes in meteorites preserve some of the most dramatic evidence for the incomplete nature of the mixing of distinct presolar materials during formation of the solar system (Clayton et al. 1976). Such 'anomalies' are present in the isotopic distributions of light elements such as H, C, N, and O. Thus, some of the isotopic heterogeneities of 'primitive' solar system materials reflect the preservation of unique clues to processes occurring during formation of the solar system. We were interested in oxygen isotopes. Oxygen has three naturally occurring isotopes: ^{16}O, ^{17}O, and ^{18}O. The most abundant is ^{16}O, with a small percentage of ^{18}O and an even smaller percentage of ^{17}O. Oxygen isotope analysis considers only the ratio of ^{18}O (or ^{17}O) to ^{16}O present in a sample. Figure 3.5 is a standard three-isotope plot showing the measured ratios in three silicates (pyroxene, forsteritic olivine and nepheline) and one oxide (spinel) present in the Florence sample. Terrestrial rocks and water fall on a line of slope ~0.5 (the terrestrial fractionation line; TF). Anhydrous

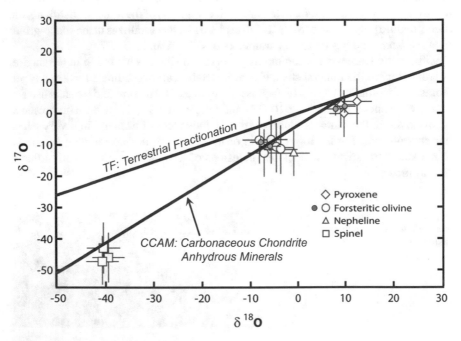

Fig. 3.5 Graph showing the oxygen three-isotope composition as suggested by Clayton et al. (1976) to discriminate terrestrial and extra-terrestrial minerals, containing data for four minerals (pyroxene, nepheline, forsteritic olivine, and spinel) in the sample studied. Error bars are 2σ. The plotted compositional indices, $\delta^{18}O$ and $\delta^{17}O$, are respectively the difference between the ratios $^{18}O/^{16}O$ and $^{17}O/^{16}O$ and the same ratios in Vienna Standard Mean Ocean Water (VSMOW), expressed in parts per mil. Terrestrial minerals fall along the upper gently sloping line (TF); the oxygen isotope compositions measured for our sample lie along the CCAM line corresponding to anhydrous minerals in the CO or CV carbonaceous chondrites. Modified after Bindi et al. (2012)

minerals from carbonaceous chondrite meteorites (chondrule minerals and calcium-aluminum-rich inclusions, or CAIs) fall on a line of slope ~0.94 that extends to very ^{16}O-rich compositions in the case of CAIs. This line is referred to as the carbonaceous chondrite anhydrous mineral (CCAM) line.

The analyses of the four minerals from the Florence sample fall along the CCAM line and are very far from the TF line. Noteworthy, spinel data (empty squares) fall in the area usually occupied by oxide minerals coming from calcium-aluminum-rich inclusions. The other extraterrestrial objects plotting near such compositions are the Solar wind (McKeegan et al. 2011) and amoeboid olivine aggregates (Krot et al. 2002). The main conclusion of this isotope study is that the silicates and oxides in the Florence sample are unquestionably extraterrestrial in origin. More in detail, the measured isotope values are close to those usually observed for CAIs from CV3 and CO3 carbonaceous chondrites, the oldest meteorites known in our Solar System.

The discovery was a breakthrough again: it indicated new processes important in forming the initial matter that came together to form the cores of planets. Confirming the origin of the sample and finding more samples to study suddenly took on new scientific importance.

3.7 A Conceivable Explanation for Their Formation: A Hypervelocity Impact in Outer Space

Earlier studies by Rubin (1994) suggested that ordinary chondrites containing Cu-bearing FeNi grains underwent impacts that led to shock-generated melting and the extraction of the Cu into metallic droplets. In the meteorite containing natural quasicrystals we have observed that FeNi can host Al and Cu, both separately and together (Hollister et al. 2014); it is, therefore, reasonable to suppose that these Al- and Cu-bearing FeNi phases were the initial source of the CuAl metals. This could explain how the lithophilic Al and chalcophylic Cu became associated and may account for the very low oxidation conditions implied by the presence of metallic Al and Cu (Bindi et al. 2012).

The simplest hypothesis, then, is that the shock locally generated the heat and pressure necessary to extract the Al and Cu from the FeNi metals and to initiate the local melting of metals and silicate; the Al and Cu would then be dissolved in a silicate melt from which the Al–Cu–Fe alloys precipitated. The local, transient ultra-high pressures led to the nucleation and growth of high-pressure phases like stishovite and ringwoodite (Hollister et al. 2014). Icosahedrite could have grown from melted metals, and stishovite could have grown around it while the pressure was still high, thus explaining the stishovite grains with icosahedrite inclusions found in the original museum sample (Fig. 3.4). A plausible alternative hypothesis is that the Cu–Al metals formed in some nebular process that removed the Cu and Al from FeNi *prior* to the impact and that the impact resulted in the remelting, crystallization and rapid cooling of the CuAl metals. This hypothesis can account for the icosahedrite inclusions in stishovite and also explain why some of the CuAl metal grains do not show evidence of shock. The reducing conditions required to make the metal are consistent with those inferred to exist in the presolar nebula at 4.5 Gya (Bindi et al. 2012). A further implication of a nebular origin of the metal is that the unshocked CuAl metals in general and the quasicrystals in particular date to the beginnings of the Solar System. Both hypotheses leave open the question of how the Al- and Cu-bearing FeNi phases formed from the solar nebula in the first place, and motivated laboratory-based static high-pressure and shock-experiments (see next chapters) to test the notion that shock-melting leads to the extraction of Al and Cu from FeNi phases.

3.8   The Meteorite Parent Body

To better understand the origin of these exotic phases, and the relationship of Khatyrka to other CV chondrites, we also measured the He and Ne content in six individual, ~40-μm-sized olivine grains from the meteorite (Meier et al. 2018). The measurement of the noble gases allows them to be dated using the method of age determination that depends on the production of helium during the decay of the radioactive isotopes ^{235}U, ^{238}U and ^{232}Th. Because of this decay, the helium content of a mineral capable of retaining helium will increase during the lifetime of that mineral, and the ratio of helium to its radioactive progenitors can be used as a measure of geologic time. If the parent isotopes are measured, the noble-gas dating method is referred to as Uranium,Thorium–Helium dating (U,Th–He dating).

We found a cosmic-ray exposure age—the time between when a meteoroid was broken off its parent body (such as an asteroid) and its arrival on Earth as a meteorite—of about 2–4 Ma (if the meteoroid was <3 m in diameter, more if it was larger). The ages obtained by means of the U,Th–He dating for the olivine grains confirmed that the meteorite containing quasicrystals experienced a relatively recent (<600 Ma) shock event that created pressure and temperature conditions sufficient to form both the quasicrystals and the high-pressure phases found in the meteorite. Meier et al. (2018) proposed that the parent body of the meteorite is the large asteroid 89 Julia (a main-belt asteroid discovered in 1866), based on: (i) its peculiar, but matching reflectance spectrum, (ii) characteristic evidence for an impact/shock event within the last few 100 Ma (which formed the Julia family), and (iii) its location close to strong orbital resonances, so that the meteoroid could plausibly have reached Earth within its rather short cosmic-ray exposure age. Furthermore, 89 Julia is a K-type asteroid, a type of relatively uncommon asteroids with a typical spectrum resembling that of CV-carbonaceous chondrites as is Khatyrka.

3.9 A Journey to the End of the World

The only way to find more sample for additional studies was to return to where the original was found, one of the most remote places on the planet. Despite overwhelming barriers, on July 22, 2011, a team of ten scientists from the US, Russia and Italy, two drivers and a cook who is also a lawyer (Fig. 3.6) gathered at the edge of the town of Anadyr, the capital of Chukotka, ready to board the odd-looking double-track vehicles that would take them across the tundra and into the Koryak Mountains to the Listvenitovyi stream, 230 km to the southwest.

As the team boarded the two strange vehicles with great hopes, there was no telling if they would find anything. Some on the expedition and many at home maintained doubts about whether there was any truth to the entire story of the origin of the quasicrystal-bearing rock.

The three-week adventure included broken axles, fires, bears, mosquitoes and August snowstorms. At the Listvenitovyi stream, one-and-a-half tons of clay and sediment were dug by hand and painstakingly panned in the same manner that gold was panned in the California gold rush. The results of this hard labor would not

Fig. 3.6 Members of the Koryak expedition team (left to right): Bogdan Makovskii (driver), Vadim Distler (IGEM, Russia), Marina Yudovskaya (IGEM, Russia), Valery Kryachko (Voronezh, IGEM), Glenn MacPherson (Smithsonian Institution, USA), Luca Bindi (University of Firenze, Italy), Victor Komelkov (driver), Olga Komelkova (cook), Alexander Kostin (BHP Billiton, USA), Christopher Andronicos (Purdue, USA), Michael Eddy (Purdue, USA), Will Steinhardt (Harvard, USA). At the center is Paul Steinhardt (Princeton, USA), leader of the expedition. Photo by W. M. Steinhardt

be known until weeks after the journey was over and the samples were studied grain-by-grain in the laboratory.

Over the next five years, ten new samples were found through a painstaking grain-by-grain search that have firmly established that the quasicrystal and the rock containing it are definitely part of a carbonaceous chondritic meteorite with calcium-aluminum inclusions that date back 4.5 Gya to the formation of the Solar System (Lin et al. 2017 and references therein). The new meteorite find has been named Khatyrka (MacPherson et al. 2013). The name derives from the Khatyrka river, which is one of the main rivers draining the Koryak Mountains. That river is also the namesake of the mineral, khatyrkite, which gives an added symmetry to the meteorite name. Khatyrka has been approved by the Nomenclature Committee of the Meteoritical Society, and representative specimens are on deposit at the U.S. National Museum of Natural History, Smithsonian Institution, Washington D.C.

3.10   New Samples and New Quasicrystals

The Khatyrka meteoritic fragments recovered from the expedition presented a range of evidence indicating that an impact shock generated a heterogeneous distribution of pressures and temperatures in which some portions of the meteorite reached at least 10 GPa and 1200 °C (Hollister et al. 2014). Among the Khatyrka fragments, the second natural discovered quasicrystal has been found (Bindi et al. 2015a). It was found as small grains, one of which is in contact with a $(Fe,Mg)_2SiO_4$ phase (marked as "olivine" in Fig. 3.7). This is either an intermediate composition olivine or the high-pressure polymorph ahrensite (olivine with the spinel structure), which was also observed in another Khatyrka grain (Hollister et al. 2014). The new quasicrystal has composition $Al_{71}Ni_{24}Fe_5$ and is the first known natural quasicrystal with decagonal symmetry, a periodic stacking of layers containing quasiperiodic atomic arrangements with ten-fold symmetry (Fig. 3.8). The mineral and its name, decagonite, have been approved by the Commission on New Minerals, Nomenclature and Classification of the International Mineralogical Association (Bindi et al. 2015b).

As already observed for icosahedrite (Bindi et al. 2009, 2011), decagonite exhibits a high degree of structural perfection, particularly the absence of significant phason strains (Fig. 3.8). This is highly unusual because such a high degree of perfection was obtained in a quasicrystal intergrown with other phases (including high-pressure minerals) under conditions that are clearly far from equilibrium and not under controlled laboratory conditions. We think that either the mineral samples formed without phason strain in the first place, or subsequent annealing was sufficient for phason strains to relax away.

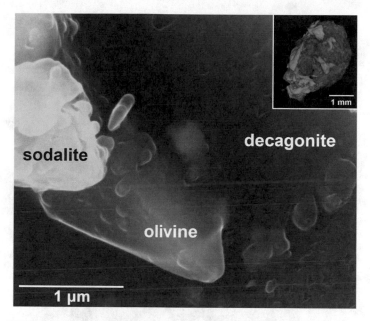

Fig. 3.7 SEM-BSE image of decagonite in apparent growth contact with olivine and sodalite. A micro CT-SCAN 3D-image of the whole sample containing decagonite is shown in the inset. The brighter and the darker regions are Cu–Al metals and meteoritic silicates, respectively. Modified after Bindi et al. (2015b)

3.11 Something Unexpected

Among the numerous new minerals described in the Khatyrka meteorite (steinhardtite, AlNiFe, Bindi et al. 2014; stolperite, AlCu, hollisterite Al_3Fe and kryachkoite, $(Al,Cu)_6(FeCu)$, Ma et al. 2017; proxidecagonite, $Al_{34}Ni_9Fe_2$, Bindi et al. 2018), particularly notable was the description of the possible third natural quasicrystal (Bindi et al. 2016), which coexists with icosahedrite in the same fragment (Fig. 3.9). We will refer at it as i-phase *II* to distinguish it from i-phase *I* (icosahedrite) with the composition of $Al_{63}Cu_{24}Fe_{13}$.

The new icosahedral phase has a composition $Al_{61.0(8)}Cu_{32.2(8)}Fe_{6.8(4)}$, which is outside (Fig. 3.10) the measured equilibrium stability field at standard pressure of the previously reported Al–Cu–Fe quasicrystal ($Al_xCu_yFe_z$, with x between 61 and 64%, y between 24 and 26%, z between 12 and 13%).

A composition close to that of i-phase *II* was reported by Zhang and Lück (2003) and Zhang et al. (2005) during investigations on the Al–Cu–Fe system with low Fe content starting from an alloy with composition $Al_{56.8}Cu_{37}Fe_{6.2}$ and annealed at 660 °C. Based on scanning electron microscopy and X-ray powder diffraction measurements, they claimed an icosahedral phase with composition $Al_{62.3}Cu_{28.6}Fe_{9.1}$, with significantly higher percentage of Fe and lower Cu than observed here in i-phase *II*.

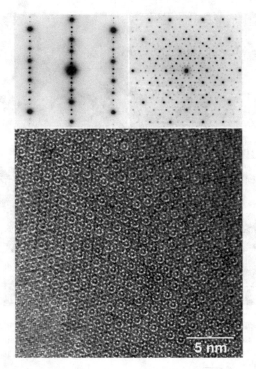

Fig. 3.8 The bottom panel is a high-resolution transmission electron microscopy (HRTEM) image showing that the real-space structure of decagonite consists of a homogeneous, quasiperiodic and ten-fold symmetric pattern. The top panels report two selected-area electron diffraction patterns collected down the ten-fold axis (right) and along an axis perpendicular to the ten-fold plane (left). The combination of quasiperiodicity (ten-fold symmetry) in one plane and periodicity along the third dimension is characteristic of decagonal quasicrystals

The possible reason of the fact that i-phase *II* has been not yet found among the products of laboratory experiments could be linked to a kinetical stabilization of the phase, mainly due to the supercooling experienced by our meteorite. This would imply that i-phase *II* is not a thermodynamically stable phase at any range of pressure and temperature. However, if we look at the Khatyrka grain in Fig. 3.9, different portions containing metallic alloys, including quasicrystals, can be observed (dashed white rectangle in Fig. 3.9). In particular, there are regions containing icosahedrite and others, which exhibit a similar texture, with i-phase *II* (see bottom panels of Fig. 3.9). The similarity in textures of the different regions reflects similar cooling conditions, also in agreement with what expected from the Al–Cu–Fe phase diagram. So, to have metallic regions with icosahedrite or i-phase *II* seems to be related to a different starting composition. In detail, i-phase *II* could have started to nucleate in a region of the Al–Cu–Fe phase diagram well outside that usually considered as stable for the icosahedral phase. The high pressure and temperature obtained during a shock could have induced an expansion of the stability field of the i-phase. In this case, i-phase *II* would represent an independent new high-pressure phase.

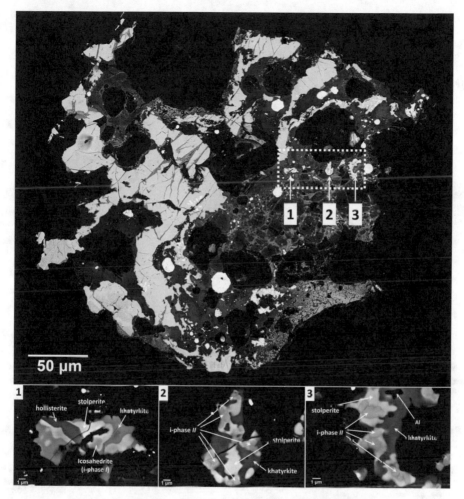

Fig. 3.9 Backscattered electron images of the grain 126A of the Khatyrka meteorite. White dashed box indicates the region to be enlarged in the bottom panels. Panels 1, 2 and 3 show the different associations of minerals in the three metal assemblages. Modified after Bindi et al. (2016)

The discovery of i-phase *II* is outstanding for several motives. Indeed, not only it is the first recognized example of the coexistence of two different Al–Cu–Fe icosahedral phases, but it represents the first case of an unknown chemical composition able to give quasicrystals found in nature prior to being discovered in the laboratory.

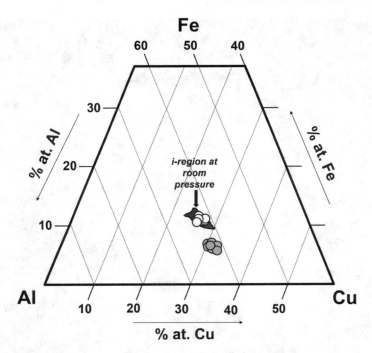

Fig. 3.10 Subsolidus projection of the ternary Al–Cu–Fe phase diagram in the vicinity of the icosahedral phase (modified after Bancel 1991). The maximal extent of the icosahedral phase occurs within the gray area labelled i-region. White and light gray circles correspond to data from icosahedrite and i-phase *II*, respectively. Errors are within the size of the symbols

References

Bancel PA (1991) Order and disorder in icosahedral alloys. In: DiVincenzo DP, Steinhardt PJ (eds) Quasicrystals. Series on directions in condensed matter physics, vol 16. World Scientific, pp 17–55

Bindi L, Steinhardt PJ (2014) The quest for forbidden crystals. Mineral Mag 78:467–482

Bindi L, Steinhardt PJ, Yao N, Lu PJ (2009) Natural quasicrystals. Science 324:1306–1309

Bindi L, Steinhardt PJ, Yao N, Lu PJ (2011) Icosahedrite, $Al_{63}Cu_{24}Fe_{13}$, the first natural quasicrystal. Am Mineral 96:928–931

Bindi L, Eiler J, Guan Y, Hollister LS, MacPherson GJ, Steinhardt PJ, Yao N (2012) Evidence for the extra-terrestrial origin of a natural quasicrystal. Proc Natl Acad Sci USA 109:1396–1401

Bindi L, Yao N, Lin C, Hollister LS, Poirier GR, Andronicos CL, MacPherson GJ, Distler VV, Eddy MP, Kostin A, Kryachko V, Steinhardt WM, Yudovskaya M (2014) Steinhardtite, a new body-centered-cubic allotropic form of aluminum from the Khatyrka CV3 carbonaceous chondrite. Am Mineral 99:2433–2436

Bindi L, Yao N, Lin C, Hollister LS, Andronicos CL, Distler VV, Eddy MP, Kostin A, Kryachko V, MacPherson GJ, Steinhardt WM, Yudovskaya M, Steinhardt PJ (2015a) Natural quasicrystal with decagonal symmetry. Sci Rep 5:9111

Bindi L, Yao N, Lin C, Hollister LS, Andronicos CL, Distler VV, Eddy MP, Kostin A, Kryachko V, MacPherson GJ, Steinhardt WM, Yudovskaya M, Steinhardt PJ (2015b) Decagonite, $Al_{71}Ni_{24}Fe_5$, a quasicrystal with decagonal symmetry from the Khatyrka CV3 carbonaceous chondrite. Am Mineral 100:2340–2343

Bindi L, Lin C, Ma C, Steinhardt PJ (2016) Collisions in outer space produced an icosahedral phase in the Khatyrka meteorite never observed previously in the laboratory. Sci Rep 6:38117

Bindi L, Pham J, Steinhardt PJ (2018) Previously unknown quasicrystal periodic approximant found in space. Sci Rep 8:16271

Clayton RN, Onuma N, Mayeda TK (1976) A classification of meteorites based on oxygen isotopes. Earth Planet Sci Lett 30:10–18

Henley C (1991) In: DiVincenzo DP, Steinhardt PJ (eds) Quasicrystals: the state of the art. World Scientific, Singapore, pp 429–524

Hollister LS, Bindi L, Yao N, Poirier GR, Andronicos CL, MacPherson GJ, Lin C, Distler VV, Eddy MP, Kostin A, Kryachko V, Steinhardt WM, Yudovskaya M, Eiler JM, Guan Y, Clarke JJ, Steinhardt PJ (2014) Impact-induced shock and the formation of natural quasicrystals in the early solar system. Nat Commun 5:3040

Janot C (1994) Quasicrystals: a primer. Oxford University Press, Oxford

Krot AN, McKeegan KD, Leshin LA, MacPherson GJ, Scott ERD (2002) Existence of an ^{16}O-rich gaseous reservoir in the solar nebula. Science 295:1051–1054

Levine D, Steinhardt PJ (1984) Quasicrystals: a new class of ordered structures. Phys Rev Lett 53:2477–2480

Levine D, Lubensky TC, Ostlund S, Ramaswamy A, Steinhardt PJ (1985) Elasticity and defects in pentagonal and icosahedral quasicrystals. Phys Rev Lett 54:1520–1523

Lin C, Hollister LS, MacPherson GJ, Bindi L, Ma C, Andronicos CL, Steinhardt PJ (2017) Evidence of cross-cutting and redox reaction in Khatyrka meteorite reveals metallic-Al minerals formed in outer space. Sci Rep 7:1637

Lu PJ, Deffeyes K, Steinhardt PJ, Yao N (2001) Identifying and indexing icosahedral quasicrystals from powder diffraction patterns. Phys Rev Lett 87:275507

Lubensky TC, Ramaswamy S, Toner J (1985) Hydrodynamics of icosahedral quasicrystals. Phys Rev B Condens Matter 32:7444–7452

Lubensky TC, Socolar JES, Steinhardt PJ, Bancel PA, Heiney PA (1986) Distortion and peak broadening in quasicrystal diffraction patterns. Phys Rev Lett 57:1440–1443

Ma C, Lin C, Bindi L, Steinhardt PJ (2017) Hollisterite (Al$_3$Fe), kryachkoite (Al, Cu)$_6$(Fe, Cu), and stolperite (AlCu): three new minerals from the Khatyrka CV3 carbonaceous chondrite. Am Mineral 102:690–693

MacPherson GJ, Andronicos CL, Bindi L, Distler VV, Eddy MP, Eiler JM, Guan Y, Hollister LS, Kostin A, Kryachko V, Steinhardt WM, Yudovskaya M, Steinhardt PJ (2013) Khatyrka, a new CV3 find from the Koryak Mountains, Eastern Russia. Meteorit Planet Sci 48:1499–1514

McKeegan KD, Kallio APA, Heber VS, Jarzebinski G, Mao PII, Coath CD, Kunihiro T, Wiens RC, Nordholt JE, Moses RW Jr, Reisenfeld DB, Jurewicz AJG, Burnett DS (2011) The oxygen isotopic composition of the sun inferred from captured solar wind. Science 332:1528–1532

Meier MMM, Bindi L, Heck PR, Neander AI, Spring NH, Riebe EI, Maden C, Baur H, Steinhardt PJ, Wieler W, Busemann H (2018) Cosmic history and a candidate parent asteroid for the quasicrystal-bearing meteorite Khatyrka. Earth Planet Sci Lett 490:122–131

Razin LV, Rudashevskij NS, Vyalsov LN (1985) New natural intermetallic compounds of aluminum, copper and zinc—khatyrkite CuAl$_2$, cupalite CuAl and zinc aluminides from hyperbasites of dunite-harzburgite formation. Zap Vses Mineral Obshch 114:90–100 (in Russian)

Rubin AE (1994) Metallic copper in ordinary chondrites. Meteoritics 29:93–98

Shechtman D, Blech I, Gratias D, Cahn JW (1984) Metallic phase with long-range orientational order and no translational symmetry. Phys Rev Lett 53:1951–1953

Tsai AP, Inoue A, Masumoto T (1987) A stable quasicrystal in Al-Cu-Fe system. Jpn J Appl Phys 26:L1505

Zhang L, Lück R (2003) Phase diagram of the Al–Cu–Fe quasicrystal-forming alloy system: I. Liquidus surface and phase equilibria with liquid. Zeit Metallk 94:91–97

Zhang L, Schneider J, Lück R (2005) Phase transformations and phase stability of the AlCuFe alloys with low-Fe content. Intermetallics 13:1195–1206

Chapter 4
From Crystals to Quasicrystals: There's Plenty of Room Between Them

Beside the three natural quasicrystals, icosahedrite (Bindi et al. 2009, 2011), decagonite (Bindi et al. 2015a, b) and the new unnamed icosahedral phase (i-phase II) with composition $Al_{61}Cu_{32}Fe_7$ (Bindi et al. 2016), we discovered the first natural periodic approximant to the decagonal quasicrystal, $Al_{71}Ni_{24}Fe_5$. The approximant, with chemical formula $Al_{34}Ni_9Fe_2$, does not correspond to any previously recognized synthetic (Lemmerz et al. 1994) or natural phase. The mineral was named proxidecagonite, derived from "periodic approximant of decagonite" (from the truncated Latin word *proxĭmus* followed by the name of the quasicrystalline mineral decagonite). The new mineral and its name have been approved by the Commission on New Minerals, Nomenclature and Classification of the International Mineralogical Association (Bindi et al. 2018).

"Periodic approximant" is an accepted technical term that refers to a crystalline solid with similar chemical composition to a quasicrystal, but whose atomic arrangement is slightly distorted so that the symmetry conforms to the conventional laws of three-dimensional crystallography. Periodic approximants can be considered the missing link between quasicrystals and crystals and are very useful because they provide a well-defined starting point for models of the local atomic structure of the corresponding quasicrystals. The discovery of approximants forming at high pressure may indicate the existence and stability of yet more types of quasicrystals at non-standard conditions.

Proxidecagonite was found in one of the meteoritic fragments of the Khatyrka meteorite and crystallizes in a space group of type *Pnma* ($R_1 = 2.46\%$ for 2360 observed reflections [$F_o > 4\sigma(F_o)$ level] and 110 parameters). Rather than close-packing of the atoms, the structure can be described as a four-layer stacking along [010]. The two atomic layers at $y = 0$ and $y = 1/2$ are puckered while the others at $y = 1/4$ and $y = 3/4$ are flat. The structure can be described as a close-packing of corner-sharing of the following empty (non-centered) polyhedra: Al_6 octahedra, Ni_2Al_3 and $NiAl_4$ trigonal bipyramids, Al_7 distorted pentagonal bipyramids, and Al_5 square pyramids. In detail, Ni_1 and Ni_7 form pentagonal bipyramids (Fig. 4.1), Al_2,

© The Author(s), under exclusive license to Springer Nature Switzerland AG 2020
L. Bindi, *Natural Quasicrystals*, SpringerBriefs in Crystallography,
https://doi.org/10.1007/978-3-030-45677-1_4

Ni1 (8*d*) & **Ni**7 (8*d*) sites centering pentagonal bipyramids

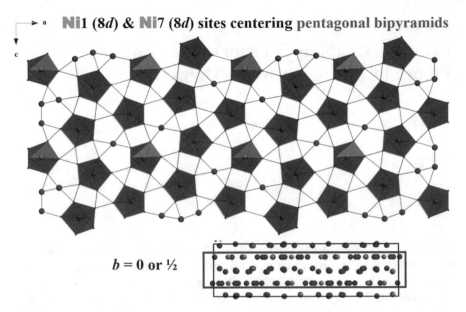

$b = 0$ or ½

Fig. 4.1 Pentagonal bypiramidal coordination environment of Ni_1 and Ni_7 in the proxidecagonite structure

Al_7 and Al_{12} cap pentagonal bipyramids (Fig. 4.2); Ni_2-to-Ni_6, Ni_8, Ni_9 and Al_1 cap trigonal bipyramids (Fig. 4.3); Al_2 and Al_{20} cap square pyramids (Fig. 4.4).

The Al_{23} site caps the longitudes of octahedra (Fig. 4.5), whereas Al_{22} caps trigonal pyramids (Fig. 4.6).

The pentagonal bipyramids centering Al (Ni_2Al_{10}) and those centering Ni (Al_7) share corners with trigonal bipyramids in the pentagonal cavities, whereas the empty distorted pentagonal bipyramids share edges with three trigonal bipyramids. Remarkably, the polyhedra with pentagonal symmetry exhibit a wave-like distribution when seen down [010], with the cavities occupied by other polyhedra arranged in a wave-like fashion as well. In this way, Ni atoms are able to create strong Ni–Al polar-covalent connections within the structure.

Interestingly, the tiling of rhombs that is usually observed in periodic approximants (Fig. 4.7) assumes similar distributions in proxidecagonite and in o'-$Al_{13}Co_4$ (Fleischer et al. 2010). The distribution in proxidecagonite differs from that observed in m-$Al_{13}Co_4$ (Grin et al. 1994) and o-$Al_{13}Co_4$ (Dolinšek et al. 2009) as shown in Fig. 4.8.

As for icosahedrite, decagonite and i-phase *II* in the Khatyrka meteorite, the natural crystalline approximant could be formed in a hypervelocity collision among asteroids in outer space. This would explain why an approximant with the composition of proxidecagonite, $Al_{34}Ni_9Fe_2$, has never been observed among synthetic products of experiments in the well-characterized Al–Ni–Fe system. If proxidecagonite is a shock-produced phase, it could be the product of a favorable kinetic pathway and not a thermodynamically stable compound. The latter still represents an open

Al6 *(4c)* & **Al**12 *(4c)* sites capping **pentagonal bipyramids**

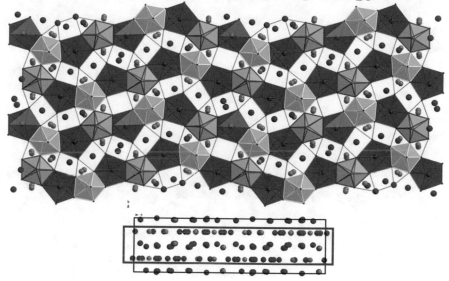

Fig. 4.2 Coordination environment of Al$_6$ and Al$_{12}$ capping pentagonal bipyramids in the proxidecagonite structure

Ni5 *(4c)* & **Al**1 *(4c)* sites capping trigonal bipyramids

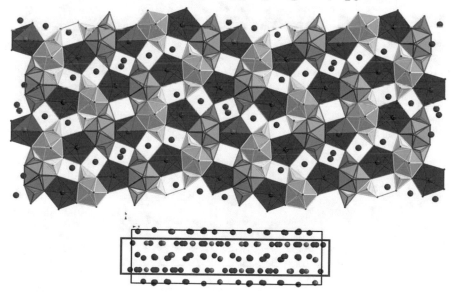

Fig. 4.3 Coordination environment of Ni$_5$ and Al$_1$ capping trigonal bipyramids in the proxidecagonite structure

Al2 (4c) sites capping square pyramids

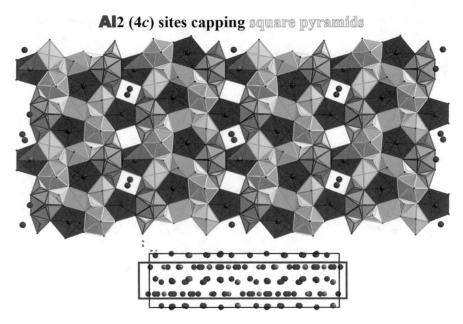

Fig. 4.4 Coordination environment of Al_2 and Al_{20} capping square pyramids in the proxidecagonite structure

Al23 (4c) sites capping longitudes of octahedra

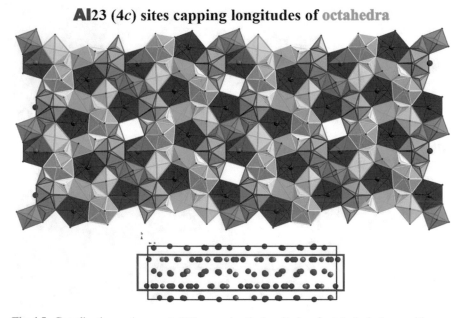

Fig. 4.5 Coordination environment of Al_{23} capping the longitudes of octahedra in the proxidecagonite structure

Al22 (4c) sites capping trigonal pyramids

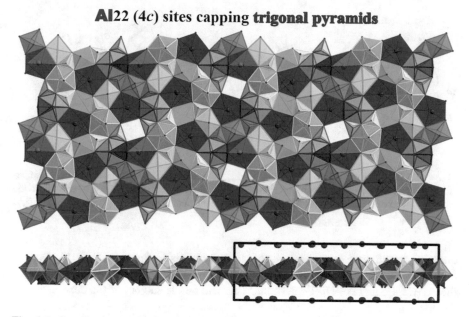

Fig. 4.6 Coordination environment of Al_{22} capping trigonal pyramids in the proxidecagonite structure

Fig. 4.7 Tiling of rhombs (red) in the proxidecagonite structure

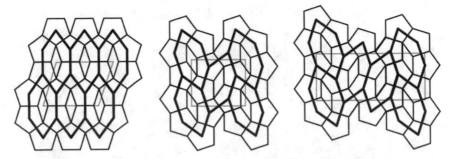

Fig. 4.8 Tilings characterizing the relationships between the closely related structures of m-$Al_{13}Co_4$ (left), o-$Al_{13}Co_4$ (middle) and proxidecagonite (right). The hexagon tilings apply to the flat layers, whereas the pentagon/rhomb tilings apply to the puckered ones. The unit cells are marked by gray lines

question especially because recent shock experiments (Asimow et al. 2016; Oppenheim et al. 2017a, b) have shown that quasicrystals can be easily produced in both the Al–Cu–Fe and Al–Ni–Fe systems. This will be the subject of the next chapters.

References

Asimow PD, Lin C, Bindi L, Ma C, Tschauner O, Hollister LS, Steinhardt PJ (2016) Shock synthesis of quasicrystals with implications for their origin in asteroid collisions. Proc Natl Acad Sci USA 113:7077–7081

Bindi L, Steinhardt PJ, Yao N, Lu PJ (2009) Natural quasicrystals. Science 324:1306–1309

Bindi L, Steinhardt PJ, Yao N, Lu PJ (2011) Icosahedrite, $Al_{63}Cu_{24}Fe_{13}$, the first natural quasicrystal. Am Mineral 96:928–931

Bindi L, Yao N, Lin C, Hollister LS, Andronicos CL, Distler VV, Eddy MP, Kostin A, Kryachko V, MacPherson GJ, Steinhardt WM, Yudovskaya M, Steinhardt PJ (2015a) Natural quasicrystal with decagonal symmetry. Sci Rep 5:9111

Bindi L, Yao N, Lin C, Hollister LS, Andronicos CL, Distler VV, Eddy MP, Kostin A, Kryachko V, MacPherson GJ, Steinhardt WM, Yudovskaya M, Steinhardt PJ (2015b) Decagonite, $Al_{71}Ni_{24}Fe_5$, a quasicrystal with decagonal symmetry from the Khatyrka CV3 carbonaceous chondrite. Am Mineral 100:2340–2343

Bindi L, Lin C, Ma C, Steinhardt PJ (2016) Collisions in outer space produced an icosahedral phase in the Khatyrka meteorite never observed previously in the laboratory. Sci Rep 6:38117

Bindi L, Pham J, Steinhardt PJ (2018) Previously unknown quasicrystal periodic approximant found in space. Sci Rep 8:16271

Dolinšek J, Komelj M, Jeglič P, Vrtnik S, Stanić D, Popčević P, Ivkov J, Smontara A, Jagličć Z, Gille P, Grin Y (2009) Anisotropic magnetic and transport properties of orthorhombic $Al_{13}Co_4$. Phys Rev B 79:184201

Fleischer F, Weber T, Jung DY, Steurer W (2010) o'-$Al_{13}Co_4$, a new quasicrystal approximant. J Alloys Compd 500:153–160

Grin J, Burkhardt U, Ellner M, Peters K (1994) Crystal structure of orthorhombic Co_4Al_{13}. J Alloys Compd 206:243–247

Lemmerz U, Grushko B, Freiburg C, Jansen M (1994) Study of decagonal quasicrystalline phase formation in the Al-Ni-Fe alloy system. Philos Mag Lett 69:141–146

Oppenheim J, Ma C, Hu J, Bindi L, Steinhardt PJ, Asimow PD (2017a) Shock synthesis of five-component icosahedral quasicrystals. Sci Rep 7:15629

Oppenheim J, Ma C, Hu J, Bindi L, Steinhardt PJ, Asimow PD (2017b) Shock synthesis of decagonal quasicrystals. Sci Rep 7:15628

Chapter 5
High Pressure Needed! The Crystallography of Quasicrystals at Extreme Conditions

It was shown in the previous chapters that the Khatyrka meteorite presented a range of evidence indicating that an impact shock generated a heterogeneous distribution of pressures and temperatures in which some portions of the meteorite reached at least 8–10 GPa and 1200 °C (Hollister et al. 2014). Beside this estimation, local pressure spikes in shocked meteorites may be significantly higher in the first tens of nanoseconds before pressure equilibration is reached. It was then crucial to verify that quasicrystals could survive such extreme conditions. Furthermore, HP-HT studies could be important to investigate the effects of pressure and temperature on the kinetic and thermodynamic stability of the quasicrystal structure relative to possible isochemical crystalline or amorphous phases.

Previous studies on synthetic icosahedral Al–Cu–Fc quasicrystals showed no phase transitions over pressures up to 35 GPa (at room temperature) and over a limited temperature range of 800–1100 K at ambient pressure (Sadoc et al. 1994, 1995; Lefebvre et al. 1995). So, the questions were: What does happen at pressures above 35 GPa? And, what about simultaneous HP/HT studies?

5.1 In Situ Synchrotron X-Ray Powder Diffraction Experiments of Icosahedrite up to About 50 GPa in Both Compression and Decompression

High-pressure diamond-anvil-cell (hereafter HP-DAC) experiments using the synthetic analog of icosahedrite produced by Bancel (1991) were performed at the 16BM-D beamline, HPCAT (Advanced Photon Source, APS, Argonne National Laboratory, ANL). These experiments aimed to investigate the evolution of the icosahedral structure under pressure, the determination of the cell parameter a_{6D} (defined below) and the equation of state. A total of 50 diffraction patterns were collected during compression and decompression. Figure 5.1 shows a characteristic spotty diffraction pattern that was consistently observed during our experiments, and

© The Author(s), under exclusive license to Springer Nature Switzerland AG 2020
L. Bindi, *Natural Quasicrystals*, SpringerBriefs in Crystallography,
https://doi.org/10.1007/978-3-030-45677-1_5

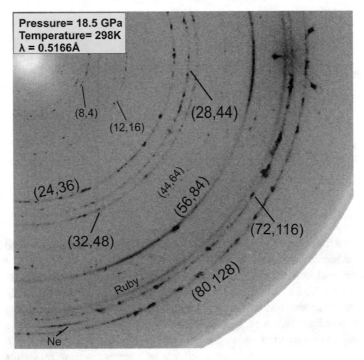

Fig. 5.1 Representative X-ray diffraction pattern of icosahedral $Al_{63}Cu_{24}Fe_{13}$ collected at $P =$ 18.5 GPa. Diffraction rings and weak single spots are typical features of the collected patterns. The Debye-Scherrer rings are labeled using the Cahn indices (N, M) (see below) described in Janot (1994). Modified after Stagno et al. (2015)

which was the result of the heterogeneous size distribution of the quasicrystal grains in the sample.

Figure 5.2 shows the variation of the d-spacings for 7 known diffraction peaks with increasing pressure and after decompressing the sample to ambient pressure. The intensities of the peaks appear to be strongly affected by the preferred orientation of the powder grains, as can be observed in Fig. 5.1. The shift in d-spacings with increasing pressure can also be seen for most of the peaks up to the target pressure. In addition, peak broadening is apparent as the pressure increases and can be attributed to an increase of the mosaic spread and local strains. After the sample was decompressed to ambient pressure, peaks were still broadened, in agreement with what was reported by Sadoc et al. (1994) for icosahedral $Al_{62}Cu_{25.5}Fe_{12.5}$.

The in situ X-ray diffraction measurements show that no peaks appear or disappear up to the target pressure of ~50 GPa, which excludes possible pressure-induced phase transformations, including amorphization, that has been found to occur for icosahedral Al–Li–Cu quasicrystals (Itie et al. 1996). The observed peak-broadening with pressure can be interpreted as arising from the increasing atomic disorder, perhaps due to residual stress, without changing the long-range quasicrystalline order. We determined the pressure dependence of the cell parameter a_{6D} up to the maximum

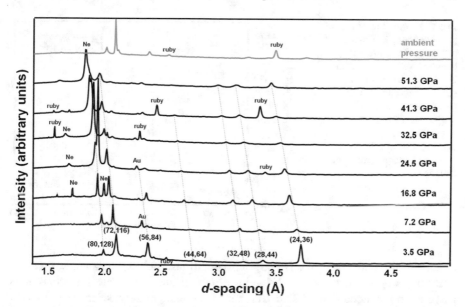

Fig. 5.2 Pressure dependence of the powder X-ray diffraction patterns of icosahedral $Al_{63}Cu_{24}Fe_{13}$ collected at room temperature in angle-dispersive mode (wavelength of 0.4246 Å). Pressure medium peaks (ruby, gold, and neon) are indicated. The diffraction peaks were indexed using Cahn indices (N, M) following the scheme described in Janot (1994; see also Steurer and Deloudi 2009). The diffraction pattern (in gray) at ambient pressure is relative to the sample after decompression. Modified after Stagno et al. (2015)

pressure of ~50 GPa. The cell parameter ("cubic" cell parameter in six-dimensional space) is defined as,

$$a_{6D} = d\sqrt{\frac{N + M\tau}{2(2 + \tau)}}$$

where d is the d-spacing in Å, N and M the Cahn indices for which the d-spacing is experimentally determined, and τ is the golden ratio, $(1 + \sqrt{5})/2$ (Steurer and Deloudi 2009). For a comprehensive definition and use of the Cahn indices see Cahn et al. (1986). The a_{6D} cell parameter is shown to gradually decrease with increasing pressure (Fig. 5.3).

The estimated reduction of the cell parameter from the ambient pressure value of 12.64 Å is about 8% as the pressure is increased to 50 GPa. In addition, the a_{6D} value at 5 GPa and room temperature is consistent with that determined by Stagno et al. (2014) under similar conditions.

The zero-pressure bulk modulus (a measurement of how resistant a substance is to compression), K_0, and its pressure derivative K_0' were determined from least-squares fits to several equation-of-state (EOS) models. In thermodynamics, an equation-of-state is a thermodynamic equation relating different state variables which describe the

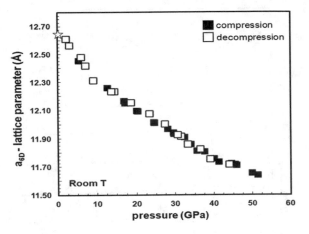

Fig. 5.3 Variation of the cell parameter a_{6D} as function of pressure: star, ambient pressure value (Bindi et al. 2011); black and white squares indicate the cell parameter determined respectively from compression and decompression experiments. Modified after Stagno et al. (2015)

state of matter under a given set of physical conditions, such as, for example, pressure and temperature. To model the behavior of matter under high-pressure conditions there are many equations-of-state that have been used in earth sciences and shock physics: e.g., Murnaghan, Birch-Murnaghan, Vinet (named for the scientists who proposed the different EOS models).

We first discuss the fit to the first order Murnaghan EOS (Angel et al. 2014), i.e. $P = K_0/K'[(V_0/V)^{K'} - 1]$, which allows direct comparison with the results from previous studies in which the experimental data were treated using the same EOS model. This resulted in $K_0 = 113.7(\pm 2.9)$ and $K'_0 = 4.22(\pm 0.22)$, respectively. It can be seen from Fig. 5.4 that K_0 and K'_0 obtained from our data are significantly lower than those obtained for $Al_{62}Cu_{25.5}Fe_{12.5}$ from previous authors using the same EOS, i.e. $139(\pm 6)$ GPa and 2.7 (Sadoc et al. 1994), and $155(\pm 10)$ GPa and 2 (Lefebvre et al. 1995).

The values obtained for our material are also diverse from those determined for the periodic $Al_{64}Cu_{24}Fe_{12}$ approximant phase (chemistry close to icosahedrite), with K_0 and K'_0 of $175(\pm 16)$ GPa and 2.00 (Lefebvre et al. 1995).

Our diffraction data were also fitted using either a third-order Birch-Murnaghan EOS, i.e. $P = 3K_0 f_E(1 + 2f_E)^{5/2}\{1 + 3/2(K' - 4)f_E + 3/2[K_0 K'' + (K' - 4)(K' - 3) + 35/9]f_E^2\}$, getting $K_0 = 110.4(\pm 2.9)$ and $K'_0 = 4.79(\pm 0.28)$, and a Vinet et al. EOS, $P = 3K_0[(1 - f_V)/f_V^2]\exp[3/2(K' - 1)(1 - f_V)]$, getting $K_0 = 109.4(\pm 2.9)$ and $K'_0 = 5.06(\pm 0.29)$. The new values, quite close to those obtained using the Murnaghan model, are in keeping with a lower bulk modulus of synthetic i-AlCuFe than those reported in literature for similar compositions.

To make a comparison with the compressional behavior of single metals (found to be stable up to $P > 100$ GPa) and that of the synthetic analogue of icosahedrite used in our experiments (Fig. 5.4), we studied the behavior of pure *fcc*-Al, *fcc*-Cu (Dewaele et al. 2004) and *hcp*-Fe (Mao et al. 1990)—as determined using the Vinet et al. and Birch-Murnaghan models. Surprisingly, synthetic i-AlCuFe shows

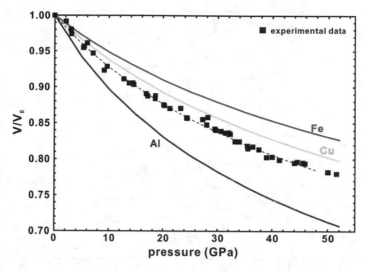

Fig. 5.4 Pressure-volume relations for icosahedrite with experimental data (black squares) fitted using the Murnaghan EOS (dashed black line). Results from Birch-Murnaghan and Vinet et al. EOS fits are here omitted since both closely overlap the Murnaghan EOS fit. Uncertainties are within the symbol size. The EOS of pure Al, Cu (Dewaele et al. 2004) and Fe (Mao et al. 1990) are also reported. Modified after Stagno et al. (2015)

a compressional behavior more similar to that of pure Cu than Al, which represents the main component of the alloy.

5.2 Double-Sided Laser-Heated DAC Experiments with In Situ Synchrotron X-Ray Diffraction at 42 GPa and up to About 2000 K

High-pressure laser heating synchrotron powder X-ray diffraction experiments were carried out at the 13ID-D beamline, GSECARS (APS, ANL) using the synthetic analogue of icosahedrite. The material was first compressed to ~42 GPa, then heated to ~1830 K. During the heating X-ray powder diffraction patterns were collected to check for possible phase transition or melting. The sample was then cooled down to 1000 K before the quenching. Looking at Fig. 5.5, it is evident that there is a decrease of the intensity of several diffraction peaks around 1560 K together with the appearance of new peaks (at d-spacings between 1.8 and 2.0 Å) belonging to the quasicrystalline structure.

Noteworthy, several strong diffraction peaks belonging to the quasicrystal structure re-appear while decreasing the temperature at ~1500 K. The peaks remain visible even after quenching the sample to room temperature. The most important conclusion is that we did not observe any amorphization or phase transformation. The

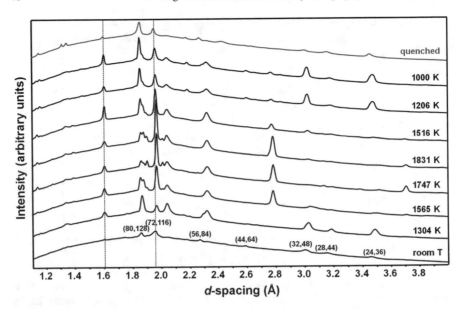

Fig. 5.5 Selected X-ray diffraction patterns for synthetic i-AlCuFe as function of temperature at P ~ 42 GPa. Diffraction peaks are indexed as described above. Vertical dashed lines indicate the 111 and 200 peaks of the Ne pressure medium. Modified after Stagno et al. (2015)

appearance of the new peaks is indeed explained by a strong preferential orientation of the quasicrystalline sample in the diamond anvil cell. Nevertheless, it is important to remark that with increasing temperature the possible formation of crystal approximants via a reversible process cannot be ruled out. More details about these experiments are given by Stagno et al. (2015).

5.3 Energy-Dispersive X-Ray Diffraction Experiments at 5 GPa and up to 1673 K

In situ energy-dispersive X-ray experiments were carried out by loading an amount of powder of synthetic i-AlCuFe in a graphite capsule directly pressurized to P ~ 5 GPa at room temperature. The X-ray diffraction patterns were then collected from 298 to 1673 K at constant press ram load of 3×10^6 N. Figure 5.6 shows the patterns in the energy range of 30–145 keV collected while increasing temperature at constant pressure. To determine the thermal dependence of interplanar distances between 1.2 and 3.4 Å, we used a low angle (3.9775°).

A general observed feature was the sharpening of the diffraction peaks with the increase of temperature. A similar behavior was observed during annealing or rapid quenching experiments of icosahedral AlCuFe-quasicrystals (Bancel 1991), and explained as likely due to the increase in quasiperiodic translational order.

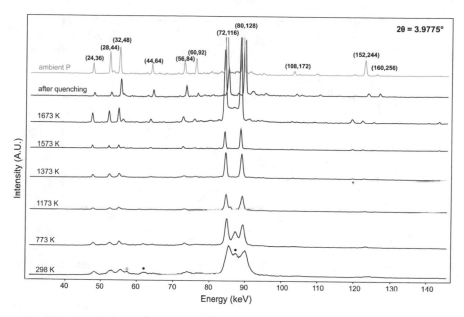

Fig. 5.6 Temperature dependence of XRD patterns collected in energy-dispersive mode for synthetic i-AlCuFe. Modified after Stagno et al. (2014)

The d-spacings, indexed using Cahn indices (N, M), decrease by 1.0–1.6% with respect to the initial values when pressure increases from ambient to 5 GPa. The temperature dependence of the observed d spacing from selected diffraction peaks is shown in Fig. 5.7. No significant change occurs during heating aside from the d-spacing increasing linearly.

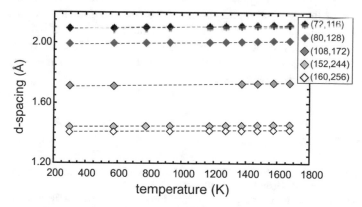

Fig. 5.7 Variation of selected d-spacings of synthetic icosahedrite with temperature. Dashed lines are linear fits. Modified after Stagno et al. (2014)

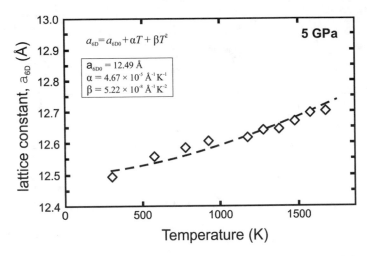

Fig. 5.8 Thermal dependence of the six-dimensional cell parameter a_{6D} for synthetic icosahedrite. Modified after Stagno et al. (2014)

More importantly, this confirms the structural stability of icosahedrite up to 1673 K. Calculations of a_{6D} at 298 K and 5 GPa gives 12.49 Å. This value is smaller than the value provided by Bindi et al. (2011) for icosahedrite (12.64 Å) and reflects the pressure effect on the cell parameter. A plot of the six-dimensional cell parameter a_{6D} with temperature is shown in Fig. 5.8.

This data can be fitted using a second-order polynomial, $a_{6D} = a_{6D0} + \alpha T + \beta T^2$, where $a_{6D0} = 12.49$ Å, T is the temperature in Kelvin, and α and β are the thermal expansion coefficients. Interestingly, the fit to our data using the values of α and β obtained by Quivy et al. (1994) for $Al_{62}Cu_{25.5}Fe_{12.5}$ (4.67×10^{-5} Å^{-1} K^{-1} and 5.22×10^{-8} Å^{-1} K^{-2}, respectively) is excellent. This could indicate that the effects of the composition of the material on the thermal expansion coefficients are small or that there is a negligible pressure effect on the volumetric expansion of i-AlCuFe. More generally, our results show that, as pressure increases, the structure of icosahedral quasicrystals remains kinetically stable (and perhaps thermodynamically stable) at higher temperatures than previously measured.

References

Angel RJ, Gonzalez-Platas J, Alvaro M (2014) EosFit7c and a Fortran module (library) for equation of state calculations. Z Kristallogr 229:405–419

Bancel PA (1991) Order and disorder in icosahedral alloys. In: DiVincenzo DP, Steinhardt PJ (eds) Quasicrystals. Series on directions in condensed matter physics, vol 16. World Scientific, pp 17–55

Bindi L, Steinhardt PJ, Yao N, Lu PJ (2011) Icosahedrite, $Al_{63}Cu_{24}Fe_{13}$, the first natural quasicrystal. Am Mineral 96:928–931

Cahn J, Shechtman D, Gratias D (1986) Indexing of icosahedral quasiperiodic crystals. J Mater Res 1:13–26

Dewaele A, Loubeyre P, Mezouar M (2004) Equations of state of six metals above 94 GPa. Phys Rev B 70:094112

Hollister LS, Bindi L, Yao N, Poirier GR, Andronicos CL, MacPherson GJ, Lin C, Distler VV, Eddy MP, Kostin A, Kryachko V, Steinhardt WM, Yudovskaya M, Eiler JM, Guan Y, Clarke JJ, Steinhardt PJ (2014) Impact-induced shock and the formation of natural quasicrystals in the early solar system. Nature Comm 5:3040

Itie JP, Lefebvre S, Sadoc A, Capitan MJ, Bessiere M, Calvayrac Y, Polian A (1996) X-ray absorption and diffraction study of the stability of two quasicrystals (AlCuFe and AlCuLi) under high pressure. In: Janot C, Mosseri R (eds) Proceedings of the fifth international conference on quasicrystals. World Scientific, Singapore, p 168

Janot C (1994) Quasicrystals: a primer. Oxford University Press, Oxford

Lefebvre S, Bessiére M, Calvayrac Y, Itie JP, Polian A, Sadoc A (1995) Stability of icosahedral Al-Cu-Fe and two approximant phases under high pressure up to 35 GPa. Philos Mag B 72:101–113

Mao H-K, Wu Y, Chen LC, Shu JF, Jephcoat AP (1990) Static compression of iron to 300 GPa and $Fe_{0.8}Ni_{0.2}$ alloy to 260 GPa: Implications for composition of the core. J Geophys Res 95:21737–21742

Quivy A, Lefevbre S, Soubeyroux JL, Filhol A, Bellissent R, Ibberson RM (1994) High-resolution time-of-flight measurements of the lattice parameter and thermal expansion of the icosahedral phase $Al_{62}Cu_{25.5}Fe_{12.5}$. J Appl Crystallogr 27:1010–1014

Sadoc A, Itie JP, Polian A, Lefevbre S, Bessiére M, Calvayrac Y (1994) X-ray absorption and diffraction spectroscopy of icosahedral Al-Cu-Fe quasicrystals under high pressure. Philos Mag B 70:855–866

Sadoc A, Itie JP, Polian A, Lefevbre S, Bessiére M (1995) Quasicrystals under high pressure. Physica B 208–209:495–496

Stagno V, Bindi L, Shibazaki Y, Tange Y, Higo Y, Mao H-K, Steinhardt PJ, Fei Y (2014) Icosahedral AlCuFe quasicrystal at high pressure and temperature and its implications for the stability of icosahedrite. Sci Rep 4:5869

Stagno V, Bindi L, Park C, Tkachev S, Prakapenka VB, Mao H-K, Hemley RJ, Steinhardt PJ, Fei Y (2015) Quasicrystals at extreme conditions: the role of pressure in stabilizing icosahedral $Al_{63}Cu_{24}Fe_{13}$ at high temperature. Am Mineral 100:2412–2418

Steurer W, Deloudi S (2009) Crystallography of quasicrystals: concepts, methods and structures. Springer series in materials science, vol 126. Springer, Heidelberg

Chapter 6
Dynamic *Versus* Static Pressure: Quasicrystals and Shock Experiments

It was shown that icosahedrite exhibits a very narrow stability field in the Al–Cu–Fe ternary system in the range 700–850 °C at ambient pressure (Zhang and Lück 2003). Nevertheless, its stability field is expanded in static high-pressure conditions, as documented in quench experiments at 1400 °C and 21 GPa (Stagno et al. 2015). The occurrence of high-pressure phases as ahrensite and stishovite (Bindi et al. 2012; Hollister et al. 2014) associated with icosahedrite and decagonite in the Khatyrka meteorite, together with the static high-pressure stability of the quasicrystalline phase, are indicative of a genesis during the high-pressure pulse of a shock event. It should be pointed out, however, there are numerous differences between the static high-pressure, high-temperature conditions studied above (as described in the previous chapter) and shock-induced conditions at nominally similar peak (P, T) conditions—including heterogeneous and rapidly time-varying pressure and temperature fields (Malavergne et al. 2001; Sharp and DeCarli 2006), typical adiabatic decompression rather than temperature quench at high pressure (Tomeoka et al. 1999), and hypersonic turbulent shear flows (Stöffler et al. 1988; Potter and Ahrens 1994; Kenkmann et al. 2000). Thus, more than high-pressure static experiments, a laboratory shock experiment producing Al–Cu–Fe icosahedral quasicrystals from discrete starting materials would give significantly strong new evidence in favor of the shock-induced origin of icosahedrite and might provide information on the shock conditions experienced by the Khatyrka meteorite.

6.1 Designing the Experiment to Shock-Produce Icosahedrite

Khatyrka meteorite fragments are composed by Cu–Al rich alloys intergrown with minerals usually found in CV3 carbonaceous chondrites, including olivine and FeNi metal (MacPherson et al. 2013). Hence, to test simultaneously whether Al–Cu–Fe alloys, quasicrystalline or not, could form by shock, we designed an experiment

© The Author(s), under exclusive license to Springer Nature Switzerland AG 2020
L. Bindi, *Natural Quasicrystals*, SpringerBriefs in Crystallography,
https://doi.org/10.1007/978-3-030-45677-1_6

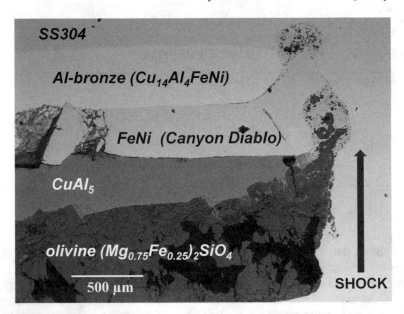

Fig. 6.1 Backscattered electron image of half of the specimen. Labels indicate the direction of shock propagation, and the layout of discs of starting material

using Cu–Al alloys in contact with both Fe^{2+}-bearing silicates and Fe-rich alloys (useful also to get information on the source of iron in icosahedrite). A stack of target discs (5 mm in diameter and 0.5–1.0 mm thick) were loaded into a cavity 5 mm below the impact surface of a stainless steel SS304 chamber with composition $Fe_{71}Cr_{18}Ni_8Mn_2Si_1$. The order of the disks along the stacking direction was: Fe-bearing forsterite with composition $(Mg_{0.75}Fe_{0.25})_2SiO_4$, $CuAl_5$ alloy, natural FeNi from the Canyon Diablo meteorite, and $Al_{14}Cu_4Fe_1Ni_1$ alloy. The target (the four disks mentioned above) was impacted by a 2 mm thick Ta flyer launched at 896 ± 4 ms^{-1}, generating a 500 ns-shock event and peak shock pressures in the various layers of the target ranging from 14 to 21 GPa. A backscattered electron image of half of the polished surface of the recovered specimen is shown in Fig. 6.1.

6.2 Shock-Produced Icosahedrite

From Fig. 6.1 it is clear that only negligible changes affect the interiors of the target layers or the interfaces in the region of one-dimensional flow near the center of the capsule (toward the right side of the image). Along the outer circumference of the chamber, however, a channel about 1 mm wide show features produced by a strong shear flow or jet with transport of material more than 1 mm down-range from its starting point. The texture exhibits angular fragments of olivine that form spherules surrounded by metal with a peculiar backscattered electron contrast (Fig. 6.2).

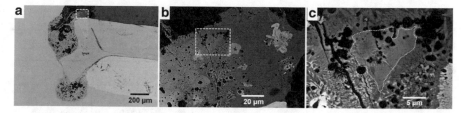

Fig. 6.2 Backscattered electron images of the recovered sample. **a** Almost the same area depicted in the center of Fig. 6.1; white dashed box indicates the region enlarged in (**b**). **b** Mixed area where $CuAl_5$ (on the right), Fe-rich metal (on the left), and olivine (at top) reacted; white dashed box indicates the region enlarged in (**c**). **c** Small icosahedral quasicrystals can be seen at higher magnification in the dotted area. Modified after Asimow et al. (2016)

Looking carefully at the boundary between the $CuAl_5$ layer and the newly formed hybrid metal region, close to the edge of the olivine layer, numerous rounded olivine grains can be observed. At the edge between the FeNi metal and $CuAl_5$, in close contact with olivine grains, we found a region (dotted area in Fig. 6.2c) that shows the presence of five-fold symmetry axes when studied by electron backscatter diffraction (EBSD) (Fig. 6.3).

Electron microprobe analyses of this grain (area 1) yield the composition reported in Table 6.1, corresponding to $Al_{72}Fe_{16}Cu_{10}Cr_1Ni_1$.

Additional grains found on the opposite side of the capsule in a comparable position with respect to the $CuAl_5$, olivine, and SS304 starting materials also showed

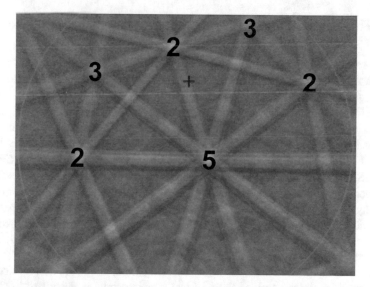

Fig. 6.3 Electron backscatter diffraction pattern of the i-phase region indicated in Fig. 6.2c, showing a combination of five-fold, three-fold and two-fold symmetry axes. Modified after Asimow et al. (2016)

Table 6.1 Electron microprobe analyses (atomic proportions on the basis of 100 atoms) for three regions of the icosahedral phase

Area	1	2	3
n	4	3	4
Al	72.3 (5)	73.3 (4)	67.94 (13)
Fe	15.7 (5)	11.3 (3)	15.42 (13)
Cu	9.7 (1)	11.57 (5)	11.30 (6)
Cr	1.02 (3)	2.73 (6)	3.73 (5)
Ni	1.23 (3)	1.11 (5)	1.62 (5)
Total	100.00	100.00	100.00

n is the number of analyses used to calculate the average

five-fold symmetry axes in their EBSD pattern and gave similar chemical compositions than those listed in Table 6.1. Noteworthy, the presence of all the five elements never occur in any of the starting materials thus indicating that the reaction forming the i-phase involved at least two of the starting materials. Given the presence of Cr and Ni, it is very likely that the i-phase formed by reaction between the stainless steel chamber and $CuAl_5$, but only close to the olivine melt spheres. The source of Fe in the i-phase has been deeply studied by successive experiments (Oppenheim et al. 2017a). The FeNi and $Al_{14}Cu_4Fe_1Ni_1$ disks seem to have not contributed to the quasicrystal formation, likely because they did not reach the required temperature to melt and react.

6.3 The Crystallography of Shock-Produced Icosahedrite

After a careful check of the polished section by scanning electron microscopy and electron microprobe, a fragment of the i-phase was handpicked and analyzed by both single-crystal and powder X-ray diffraction. The single-crystal pattern down the five-fold rotational axis, in gnomonic projection, is shown in Fig. 6.4.

The diffraction pattern exhibits the same five-fold distribution and overall symmetry as icosahedrite and also low phason strain. This is in stark contrast with what observed in metastable quasicrystals grown by rapid quenching from melt. Indeed, they possess significant phason strain that manifests as diffraction peaks that are broadened and shifted systematically from the ideal pattern (Levine et al. 1985; Lubensky et al. 1986). The diffraction peaks in the powder diffraction pattern of the shock-produced quasicrystal are all systematically shifted (Fig. 6.5) to larger *d*-spacing when compared to icosahedrite, $Al_{63}Cu_{24}Fe_{13}$ (Bindi et al. 2009, 2011).

To try to quantify the phason strain of the shock-produced quasicrystal, we calculated the second figure-of-merit test introduced by Lu et al. (2001) useful to compare the observed intensities to those of an ideal quasicrystal. As described in Chap. 3, such a figure-of-merit is defined as the logarithm of the intensity-weighted average

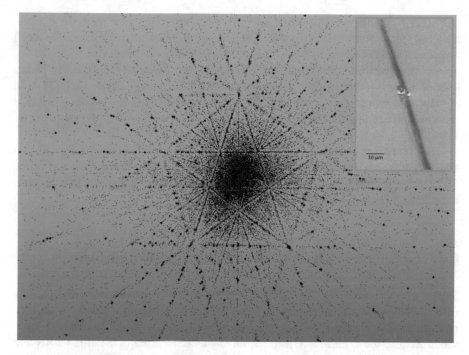

Fig. 6.4 Single-crystal X-ray diffraction pattern in gnomonic projection down the five-fold rotational axis; the extracted grain attached to a carbon fiber is shown in the inset on the right

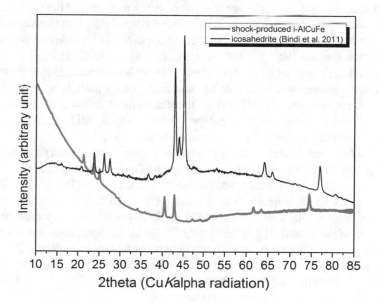

Fig. 6.5 Comparison between the powder diffraction pattern of the shock-produced quasicrystal (gray) and icosahedrite (black). Modified after Asimow et al. (2016)

$|\Delta|$, where Δ is the absolute deviation of each scattering vector from the closest-matching ideal quasicrystal peak. The intensities measured for the shock-produced phase are nearly identical to those observed for icosahedrite (Bindi et al. 2009, 2011), which scored as well as the best laboratory specimens. The analysis of the interplanar distances in the pattern gave a six-dimensional cell parameter $a_{6D} = 12.71(3)$ Å, in agreement with the value expected from the linear correlation between a_{6D} and Al/Cu ratio established in previously observed Al–Cu–Fe quasicrystals (Quiquandon et al. 1996).

The observed grain size (up to 10 μm) and low phason strain in the shock-produced quasicrystal seem inconsistent with the timescale of the high-pressure pulse, ~500 ns. More likely, quasicrystal growth and annealing continued after pressure release, even if nucleation occurred at peak pressure.

6.4 The Source of Iron in Shock-Produced Icosahedrite

With the aim to replicate the shock that affected the Khatyrka meteorite, Asimow et al. (2016) were able to synthesize five-component icosahedral quasicrystals with compositions in the range $Al_{68-73}Fe_{11-16}Cu_{10-12}Cr_{1-4}Ni_{1-2}$. As described in the previous section, the experiment used starting materials including metallic $CuAl_5$ and $(Mg_{0.75}Fe_{0.25})_2SiO_4$ Fe-bearing forsterite in a stainless steel (SS304) container. One of the main question regarding the shock-produced quasicrystals was the source of iron. Indeed, it might have originated either from the reduction of the Fe^{2+} in olivine or come directly (in the metallic state, Fe^0) from the stainless-steel recovery chamber. If the iron came from olivine, then we have to hypothesize a redox reaction with an Fe reduction process ($Fe^{2+} \rightarrow Fe^0$) compensated by the likely oxidation of Al^0 to Al^{3+} (with the consequent formation of a new generation of spinel and/or corundum). Such a redox process would be possibly associated with a thermite-type exothermic reaction. On the contrary, if the iron originated from the stainless steel chamber, then there would be no redox reaction, but the synthesis would still have likely required melting or partial melting driven by some other heat source.

To shed light on the mechanism of formation of icosahedral quasicrystals by shock, we designed two new shock recovery experiments (Oppenheim et al. 2017a). The first experiment (shot S1233) juxtaposed the same olivine composition and $CuAl_5$ alloy as in the previous experiment, but, to remove the metallic iron source, everything was embedded in a tantalum liner within an SS304 outer recovery chamber. The second experiment (shot S1234) used only $CuAl_5$ put in direct contact with SS304, eliminating both ferrous iron and silicate mineral or melt nucleation sites.

In the first experiment ($CuAl_5$ and Fe^{2+}-bearing olivine in a tantalum capsule), no quasicrystals were found. On the contrary, in the experiment with only metallic starting materials, micron-sized five-component icosahedral quasicrystals with average composition $Al_{73}Cu_{12}Fe_{11}Cr_3Ni_1$ were found at the interface between $CuAl_5$ and SS304, demonstrating nucleation of quasicrystals under shock without the need for a redox reaction.

6.4.1   Additional Shock Experiments

In the two shock experiments described below (S1233 and S1234), discs of starting material 5 mm in diameter and 1 mm total thickness were loaded into a cavity 5 mm below the impact surface of an SS304 chamber. In shot S1233, a 1 mm deep well in a Ta inner screw and a Ta lid provided 1 mm of Ta on all sides. In shot S1234, a 1 mm deep counterbore in the inner SS304 screw of the recovery chamber (Willis et al. 2006) formed the sample volume, which was closed by the SS304 driver itself. The target was impacted by a 2 mm thick Ta flyer carried by a 20 mm diameter plastic sabot.

6.4.2   Shock Without Quasicrystals Formation

Although intimate mixing between Ta, $CuAl_5$, and olivine was observed, no quasicrystals were found in this shock experiment. The composition of the starting forsteritic olivine was $(Mg_{0.75}Fe_{0.25})_2SiO_4$ (i.e., Fo_{75}—where Fo means forsterite) but SEM analyses in the region of mixing show that the recovered olivine is more Fe-rich, i.e. Fo_{80-90}. The change could be linked to the fact that part of the Fe^{2+} in the olivine did undergo reduction with a consequent enrichment in Mg in the un-reacted portions.

6.4.3   Shock with Quasicrystals Formation

The general appearance of the shocked sample is shown in Fig. 6.6. X-ray maps did not show oxygen in agreement with the fact that all-metallic starting materials were used. There is thus no need of redox reactions. The most interesting areas are those between the $CuAl_5$ layer and the SS304.

The contact between $CuAl_5$ and the SS304 is sharp, with no evidence of mixing or reaction (Fig. 6.7a); however, at the corners where there is the contact between the rear and radial edges, mixing occurs with the formation of pockets containing Cu, Al and Fe in significant amounts (Fig. 6.7b). There is also another well-defined mixed layer along the lateral boundary of the sample, with a nearly constant thickness of 20 μm (Fig. 6.7c).

A careful examination of the mixing regions shows the presence of some coarser-grained areas where grains can be studied by EBSD. The area shown in Fig. 6.6b and progressively enlarged in Fig. 6.8 is labeled region of interest (ROI) 1.

The area shown in Fig. 6.7a and progressively enlarged in Fig. 6.9 is labelled ROI 2. In both these ROIs we detected several ~1 μm grains of moderate backscatter contrast surrounded by a fine dendritic Al-rich mat and occasional Fe-rich grains (Fig. 6.8b). These regions were analyzed by EBSD. Most grains were too small to

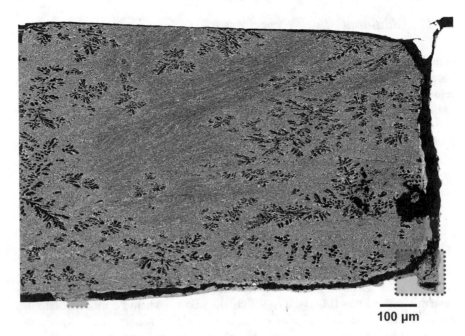

100 μm

Fig. 6.6 Backscattered electron image of shock sample S1234. The white regions around the margins are SS304. The dendritic region is CuAl$_5$. Black regions are voids where material was lost during sample preparation. Region of interest (ROI) 1, shown by the red box, is enlarged in Figs. 6.7b and 6.8. ROI2, shown by the green box, is enlarged in Figs. 6.7a and 6.9. The area along the side-wall highlighted by the blue box is enlarged in Fig. 6.7c. The inset shows a schematic (to scale; 5 mm scale bar shown) of the entire recovery chamber and flyer plate. Modified after Oppenheim et al. (2017a)

give suitable EBSD data, but many of the larger grains with similar backscatter contrast showed a pattern characterized by five-fold, three-fold and two-fold symmetry axes (Fig. 6.10). To better characterize such grains from a crystallographic point of view and to study the relationships with the surrounding phases, we used a transmission electron microscope. As concerns the Fe-rich grains (>20 atomic %), they showed cubic crystal structures in EBSD analysis, but the quality was not sufficient to distinguish the particular type of space group.

The grains in each ROI that yielded clear five-fold symmetry axes in their patterns were chemically characterized by electron microprobe analyses. The overall mean of all the chemical analyses is Al$_{73}$Cu$_{12}$Fe$_{11}$Cr$_3$Ni$_1$, on an atomic basis. The aluminum content is higher than any known stable Al–Fe–Cu ternary quasicrystals. All five elements were confidently measured above their detection limits. However, given the very small size of the grains, the chemistry was further checked using TEM analysis, which takes advantage of thin foil transmission geometry to avoid multiple scattering effects. A FIB section (i.e., a thin lamella obtained by slicing a large sample by means of the Focused Ion Beam technique) was obtained from ROI 2, as shown by the dashed white rectangle in Fig. 6.9b. The bright-field image of the FIB section shown

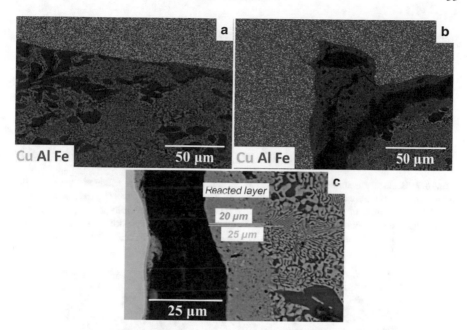

Fig. 6.7 **a** X-ray intensity map overlain on backscattered electron image (green box in Fig. 6.6). The color scheme is reported. Cu–Al alloys are blue-green. The boundary is sharp, without evidence of mixing or reaction, except for two patches at right, enlarged in Fig. 6.9 (ROI 2). **b** A similar X-ray intensity map (red box in Fig. 6.6, also shown in Fig. 6.8). Here there is a larger region of mixing and reaction (blue). **c** Enlarged back-scattered electron image of the blue box in Fig. 6.6. The bright area at far left is the SS304, whereas the black area is missing material, lost during sample preparation. The region on the right shows the typical eutectoid texture of CuAl$_5$. The 20 μm wide band down the center of the image is a fine-grained reacted zone containing Cu, Al, and Fe. Modified after Oppenheim et al. (2017a)

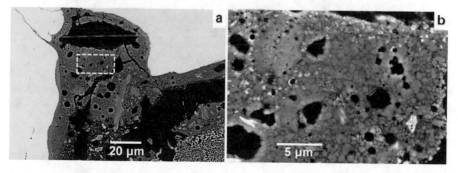

Fig. 6.8 Backscattered electron images of ROI 1, from the upper left corner of Fig. 6.6 (also shown in Fig. 6.7b). **a** Moderate magnification, with SS304 capsule, CuAl$_5$, and mixed region evident. Grains close to 1 μm in size are visible in the mixed region. **b** High magnification image of the red box area in (**a**), with voids, scattered Fe-rich grains, and abundant quasicrystal grains. Modified after Oppenheim et al. (2017a)

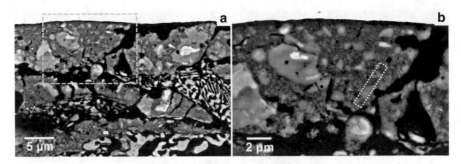

Fig. 6.9 Back-scattered electron images of Region of Interest 2. **a** Moderate magnification image shows relatively coarse grain structure with domains reaching several μm in size. **b** High-magnification image of the area shown by the red box in (**a**), with medium gray 1 μm-sized quasicrystals. The area removed by Focused Ion Beam milling for Transmission Electron Microscopy is indicated by the dashed white rectangle. Modified after Oppenheim et al. (2017a)

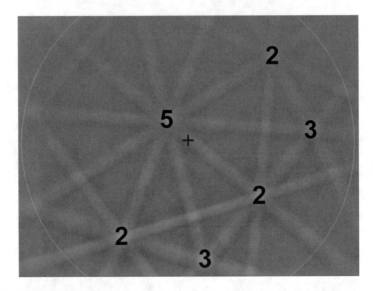

Fig. 6.10 Kikuchi pattern obtained by electron backscatter diffraction from a shock-produced icosahedral quasicrystal in ROI 2, displaying five-fold, three-fold and two-fold rotation axes. Modified after Oppenheim et al. (2017a)

in Fig. 6.11a shows numerous roughly spherical or somewhat elongated regions with variable grain sizes from 30 to 300 nm. High-magnification bright-field imaging of one of these almost spherical grains (Fig. 6.11b), exhibits dark contrast due to strong diffraction and some internal contrast. Selected-area electron diffraction (SAED) patterns reveal five-fold, three-fold, and two-fold symmetry patterns, as expected for icosahedral symmetry.

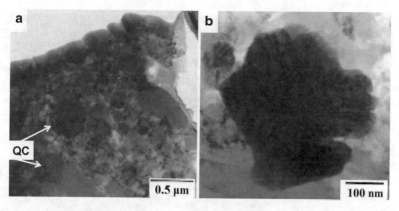

Fig. 6.11 TEM Bright-field image of the FIB section. **a** Diffraction contrast results from random orientations of individual icosahedral quasicrystals (i-QC). The arrows indicate large grains of quasicrystalline material. The matrix contains i-QCs as well as other crystalline alloys. **b** Bright-field image of an icosahedral quasicrystal grain (dark contrast) as seen down a five-fold symmetry axis. Modified after Oppenheim et al. (2017a)

For one grain, we had the possibility to rotate it and measure the angle between a five-fold and a three-fold axis (Fig. 6.12). The measured angle is ~36.7°, in excellent agreement with the ideal value of 36.38° expected for an ideal icosahedron (Steurer and Deloudi 2009).

As described above, previous authors have noted systematic relations among the composition of icosahedral quasicrystals and their lattice spacings, which can be conveniently expressed by a single cell parameter in six-dimensional space, a_{6D}. We tried to determine a_{6D} for the shock-produced quasicrystals using the lengths of the shortest vectors measurable in reciprocal space in our SAED patterns; the (8, 4) reflection, though faint, is visible in Fig. 6.12a and the (28, 44) reflection is visible in Fig. 6.12b. The (8, 4) peak gave a d-spacing of 9.053 ± 0.437 Å, corresponding to $a_{6D} = 12.804 \pm 0.619$ Å by the Cahn indices method (Cahn et al. 1986; Steurer and Deloudi 2009). Likewise, the (28, 44) peak yielded a d-spacing of 5.641 ± 0.296 Å, corresponding to $a_{6D} = 12.908 \pm 0.677$ Å. Although, these estimates of a_{6D} are fairly in agreement with the linear trend observed between a_{6D} and Al content of icosahedral quasicrystals (accurately determined by single-crystal X-ray diffraction at 12.64 ± 0.01 Å for icosahedrite with 64 at.% Al, and 12.71 ± 0.03 Å for the phase synthesized by Asimow et al. (2016) with 71 ± 2 at.% Al), the precision is actually too low to make speculations.

In the transmission electron microscope, it is straightforward to obtain an EDS spectrum from a small grain without significant interference from neighboring material. However, it is not possible to quantitatively measure Cu when the sample is mounted on a Cu TEM grid. The TEM-EDS analysis of the quasicrystals from ROI 2 gave the $Al_{83}Fe_{14}Cu_xCr_4Ni_1$ atomic proportions (the subscript x to copper indicates that it cannot be quantitatively measured as the sample is mounted on a Cu TEM grid), which are broadly consistent with the electron microprobe analyses of

Fig. 6.12 Selected-area electron diffraction patterns collected along the **a** five-fold and **b** three-fold rotation axes of a quasicrystal grain, obtained by rotating the sample by 36.7°. Selected Cahn indices are indicated (Cahn et al. 1986). **c** Inverse Fourier transform image of a high resolution TEM image of a quasicrystal as seen down a five-fold symmetry axis. Modified after Oppenheim et al. (2017a)

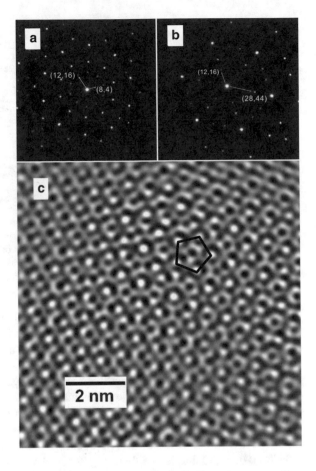

the same area. Noteworthy, Cr and Ni are definitely present in the quasicrystal and not the result of contamination from surrounding phases.

6.5 Designing the Experiment to Shock-Produce Decagonite

Icosahedrite is not the only quasicrystal found in the Khatyrka meteorite. Indeed, it contains also the first discovered decagonal quasicrystal, decagonite $Al_{74}Ni_{24}Fe_5$ (Bindi et al. 2015a, b). In the previous sections, we have described how i-quasicrystals were produced in shock experiments (mainly) at the $CuAl_5$–SS304 boundary. Here we report a new shock recovery experiment intended to reproduce Al–Ni–Fe decagonal quasicrystals (Oppenheim et al. 2017b). For this purpose, an aluminum (2024) disc was loaded into a Permalloy 80 cup (composition $Ni_{80}Fe_{15}Mo_5$) within the same stainless steel 304 (SS304) chamber (Fig. 6.13). As in the previous shock experiments, the target was impacted by a Ta flyer at 1041 ± 1 ms^{-1} and then the chamber was recovered intact, sawn open, polished, and examined with a combination of SEM, EBSD, EPMA, and TEM studies.

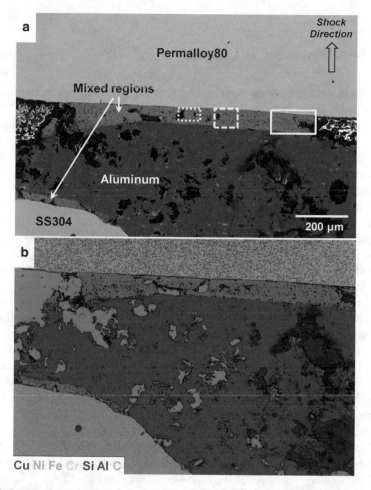

Fig. 6.13 Panoramic view of sample from shot S1235. First shock propagated from bottom towards top of the sample. **a** Backscattered electron image showing the initially 1 mm thick Al2024 layer attenuated by flow associated with impact deformation to ~0.5 mm in this area near the left side-wall of the chamber. Regions of intermediate backscatter contrast mark a reaction zone up to ~0.1 mm wide with mean atomic number intermediate between the Al-rich and Ni-rich starting materials. **b** X-ray intensity map of the same area, using the color scheme shown at lower left. Voids in the sample appear cyan due to carbon in the epoxy filler. The reddish grey regions mark the mixed layers. Modified after Oppenheim et al. (2017b)

With our surprise, decagonal quasicrystals were even easier to obtain than i-quasicrystals. At the first shock (i.e., shot S1235), we produced plentiful quasicrystalline materials having an average composition $Al_{73}Ni_{19}Fe_4Cu_2Mg_{0.6}Mo_{0.4}Mn_{0.3}$ (with minor Si and Cr) along the boundary between Al and Permalloy (Oppenheim et al. 2017b).

6.6   Shock-Produced Decagonite

A panoramic view of the shock-produced sample, extending across the full width of the Al2024 layer and attenuated by the catering flow to about half its original thickness, is shown in Fig. 6.13. Unreacted portions of the starting materials (SS304, Al2024, and Permalloy 80) together with relatively small newly-formed reacted zones (mainly in the region—about 100 μm wide—across the full rear contact between the starting materials), are clearly visible. Noteworthy, the stainless steel chamber does not appear to have contributed in any reactions.

If we look at the mixed layer at progressively higher magnification in Fig. 6.14, it becomes evident that this region contains mostly ~1 μm sized rounded grains with a few ~10 μm angular grains and numerous voids ranging from 1 to 100 μm across. As seen in previous experiments, we do not know if they are actual voids in the volume of the sample (formed upon shock-decompression) or if they formed during the preparation of the sample (i.e., polishing procedure).

All the phases in the shock-produced sample were identified and characterized by a combination of SEM-EDS and EBSD technique. The light gray grains in Fig. 6.14 were found to exhibit the 10-fold symmetry axis in their Kikuchi patterns, even if only a small fraction had the 10-fold zone axis within the cone sampled by the EBSD detector (in fact, the d-QC have a preferred orientation; see below). EDS chemical analysis pointed out the presence of seven metals above detection limits: Al, Ni, Fe, Cu, Mg, Mo, and Mn. Subsequent quantitative electron microprobe analysis shows an average composition for these regions of $Al_{73.3}Ni_{19.3}Fe_{4.3}Cu_{1.8}Mg_{0.6}Mo_{0.4}Mn_{0.3}$ (see Table 6.2).

In the mixed region of Fig. 6.14a, we also characterize several crystalline phases closely associated with the d-QC. We found cubic phases exhibiting two distinct compositions: $Al_{63}Ni_{29}Fe_5Cu_2Mo_1$ and $Al_{50}Ni_{37}Fe_8Cu_2Mo_1$ (Table 6.2; the sums of these formulas may differ from 100 due to rounding and other minor components). It was hard to determine the space-group type of the cubic phases as the two possible choices, the $Im\bar{3}m$ (A2 or *bcc*) structure (like steinhardtite) and the $Pm\bar{3}m$ (B2 or CsCl-type structure) structure, are hardly distinguishable using electron backscattered diffraction. Looking at the Al–Ni–Fe phase diagram (Chumak et al. 2007; Zhang and Du 2007), both compositions should plot in the stability field of the B2 phase. Careful examination of the EBSD patterns shows that the 111 reflection is present, which is forbidden in *bcc* and indicates that the phase is B2. This result was also confirmed by TEM (see below). Interestingly, the B2 phase and the d-QC often form a core-and-petal (respectively) flower microstructure (Fig. 6.14d).

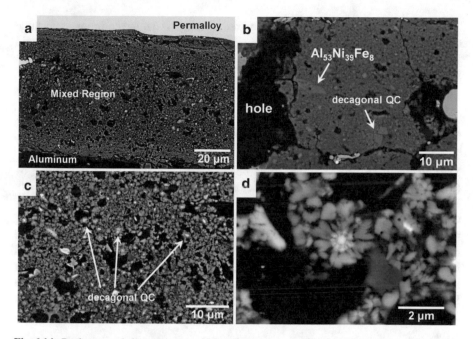

Fig. 6.14 Backscattered electron images of the mixed layer (see Fig. 6.13a for locations of frames): **a** within the ~100 μm wide mixed layer. **b** Further enlargement shows angular ~10 μm grains of a cubic phase and rounded ~1 μm grains identified as decagonal quasicrystals (d-QC). **c** The dark spots are voids, the dark gray grains are Al_9Ni_2, the light gray are d-QC, and the bright white areas are the cubic phase. **d** A "flower" with a core of cubic phase and petals of d-QC radiating outward (see text for explanations). Many of the flowers appear to display a 10-fold symmetry. Modified after Oppenheim et al. (2017b)

Table 6.2 Electron microprobe analysis (atomic proportions on the basis of 100 atoms) of the different phases

	d-QC	B2 cubic phase 1	B2 cubic phase 2	Al_9Ni_2
n	8	6	6	5
Al	73.3 (9)	63.5 (8)	50.4 (8)	80.8 (7)
Fe	4.3 (3)	4.7 (3)	7.6 (4)	3.8 (3)
Cu	1.8 (3)	2.0 (3)	2.1 (3)	0.8 (2)
Cr	n.d.	n.d.	0.14 (6)	n.d.
Ni	19.3 (7)	28.5 (7)	37.4 (9)	13.4 (4)
Mo	0.4 (1)	0.5 (1)	1.3 (1)	0.3 (1)
Mn	0.3 (2)	0.2 (2)	0.4 (2)	0.3 (2)
Mg	0.6 (1)	0.4 (1)	0.5 (1)	0.5 (6)
Si	n.d.	n.d.	0.22 (10)	n.d.

n.d. means not determined; n is the number of analyses

Al$_9$Ni$_2$ phase was also found in the mixed region. It is a quite common metastable compound in the Al–Ni–Fe system (Chumak et al. 2007) that only forms during rapid solidification. We confirmed that it possesses the monoclinic $P2_1/c$ Al$_9$Co$_2$-type structure and a Al$_{81}$Ni$_{13}$Fe$_4$Cu$_1$Mg$_1$ composition (Table 6.2).

6.7 The Crystallography of Shock-Produced Decagonite

Two FIB sections were extracted from the mixed area between the Al and Permalloy (Fig. 6.15) because it shows two distinct textures/microstructures (Fig. 6.15a).

The first microstructure is composed by d-QC, the cubic B2 phase, and Al$_9$Ni$_2$, with significant porosity. The other microstructure, closer to the Al2024 side of the reaction zone, is denser in appearance and includes predominantly fine-grained d-QC. FIB section A was taken from this latter area (Fig. 6.15b) and is shown in a bright-field TEM image in Fig. 6.15c, with three selected regions of interest highlighted with yellow rectangles (and described in the next sections). FIB section B (with two selected regions of interest highlighted with light-blue rectangles) shows a nearly uniform diffraction contrast of the d-QC grains symptomatic of a common

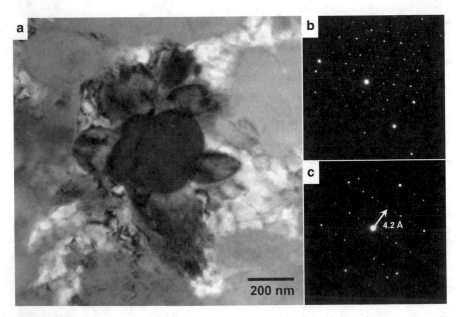

Fig. 6.15 Region of Interest 1 from FIB section A (see Fig. 6.15c). **a** Bright-field image of a decagonal quasicrystal (dark) along the 10-fold rotation zone axis. The surrounding fine grains are quasicrystals of different orientations. **b** SAED pattern of the quasicrystal along the 10-fold rotation axis. **c** SAED pattern perpendicular to the 10-fold rotation axis, showing the interlayer spacing corresponding to 4.2 Å. Modified after Oppenheim et al. (2017b)

orientation, with most of their 10-fold axes roughly parallel to the direction of shock propagation.

6.7.1 FIB Section A, Region of Interest 1

This region includes a relatively well-developed d-QC bounded by irregularly shaped QCs (Fig. 6.16). The chemical composition is roughly $Al_{66}Ni_{19}Fe_4Cu_x$ (copper is not correctly measured because the sample is held on a Cu TEM grid). Electron diffraction patterns indicate the layered nature of the structure of the decagonal quasicrystal. The spacing between layers is ~4.2 Å, when the d-QC is viewed down the 10-fold rotation axis (Fig. 6.16c). The measured value is in excellent agreement with that obtained on decagonite by single crystal X-ray diffraction (4.208(2) Å; Bindi et al. 2015b).

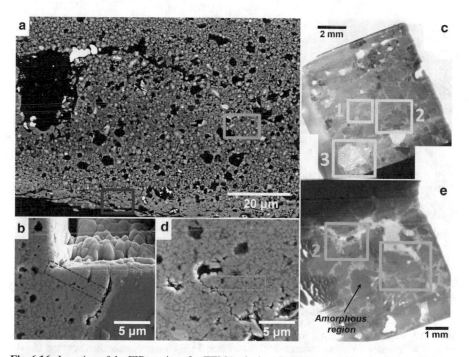

Fig. 6.16 Location of the FIB sections for TEM analysis. **a** Backscattered electron image showing texture differences across the mixed region. FIB section A was extracted from the denser-textured area (red box) close to the Al2024 layer. FIB section B was taken from an area with two prominent "flowers" (green box). **b** Secondary electron image of the area targeted for FIB section A liftout, shown by the dashed red rectangle. **c** Bright-field TEM image of extracted and thinned FIB section A. **d** Secondary electron image of the area targeted for FIB section B liftout, shown by the dashed green rectangle. **e** Bright-field TEM image of extracted and thinned FIB section B. Modified after Oppenheim et al. (2017b)

6.7.2   FIB Section A, Region of Interest 2

This region (Figs. 6.17 and 6.18) has a flower in the middle similar to those observed in the mixed region of Fig. 6.14c. The phase in the center of the flower, having moderately dark contrast in the bright-field image in Fig. 6.17 and very dark contrast in Fig. 6.18, is clearly the classic cubic B2 structure and not *bcc*-structured steinhardtite (Bindi et al. 2014).

The electron diffraction patterns collected in the central part of the flower either along the [110] (Fig. 6.17) and [111] (Fig. 6.18) zone axis always match the cubic CsCl-type (B2) structure with $a \sim 3.0$ Å. Its composition is $Al_{55}Ni_{29}Fe_4Cu_x$. It should be noted that both the diffraction patterns (Figs. 6.17 and 6.18) show the presence of superstructure reflections beside the main reflections, indicating a large periodicity in the structure. This could be due to the non-perfect stoichiometry of the phases, which could lead to some ordering of the vacancies capable of creating the observed superstructure (Yamamoto and Tsubakino 1997).

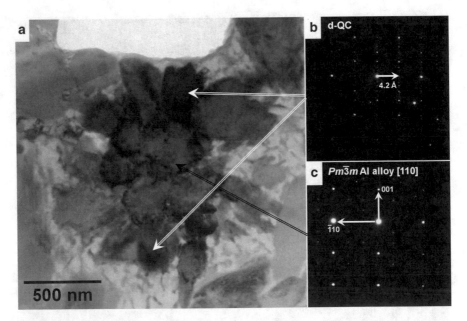

Fig. 6.17 Region of Interest 2 from FIB section A (see Fig. 6.15c). **a** Bright field image. **b** The regions with dark diffraction contrast display identical SAED images perpendicular to the 10-fold axis of the d-QC. **c** The grain at the center shows the [110] zone-axis orientation of a cubic primitive lattice (B2 structure), with weak spots in the SAED pattern indicating a superstructure. Modified after Oppenheim et al. (2017b)

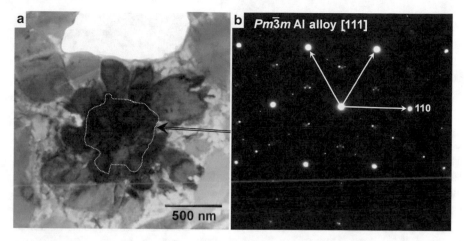

Fig. 6.18   A different orientation of ROI 2 from FIB section A. **a** Bright field image; diffraction contrast has changed due to rotation of view relative to Fig. 6.17a. **b** The zone axis of the Al-alloy at the center in this view is [111]. The intermediate weak superstructure reflections spots are indicated. Modified after Oppenheim et al. (2017b)

6.7.3   FIB Section A, Region of Interest 3

Silica-rich rounded material occurs in the FIB section A for ROI 3 (Fig. 6.19). Although some globules appear to have grain boundaries, diffraction patterns showed them to be amorphous. Silicon could come from the Permalloy starting material.

Fig. 6.19   Bright field image of region of interest 3 in FIB section A (see Fig. 6.15c). Globules of amorphous silica-rich material are visible. Modified after Oppenheim et al. (2017b)

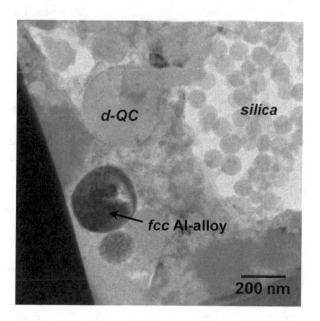

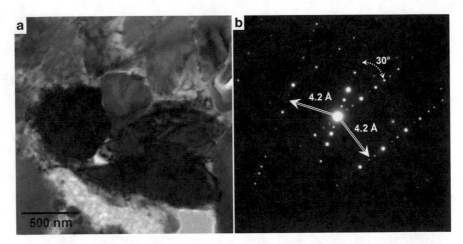

Fig. 6.20 FIB section B, Region of Interest 1 (see Fig. 6.15e). **a** Bright-field image of a cluster of d-QC grains showing similar diffraction contrast. **b** Corresponding electron diffraction pattern encompassing two of the grains, showing that both are oriented with the 10-fold zone axes in the plane of section and rotated with respect to one another by 30°. Modified after Oppenheim et al. (2017b)

6.7.4   FIB Section B, Region of Interest 1

Figure 6.20 reports a bunch of d-QC grains having their 10-fold axes in the plane of the section. Electron diffraction patterns collected on two grains down the 10-fold symmetry axis showed that they both have the common 4.2 Å layer spacing and the same forbidden axis rotated with respect to one another by 30°.

6.7.5   FIB Section B, Region of Interest 2

Figure 6.21 reports ROI 2 that contains mottled regions of *fcc* aluminum interstitial to the d-QC grains. Electron diffraction patterns of the *fcc* Al exhibit doubled spots, consistent with twinning being responsible for the mottled contrast. Some interstitial unreacted aluminum is present together with d-QC, B2, and Al_9Ni_2.

6.8   Crystallographic Remarks on Shock-Produced Decagonite

The shock-produced Al–Ni–Fe quasicrystal shows close similarities with decagonite found in the Khatyrka meteorite (Bindi et al. 2015a, b). Neglecting the minor elements found in the shock-produced synthetic quasicrystal (Cu, Mg, Mo, and Mn), the

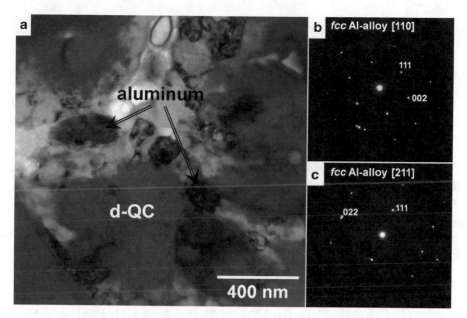

Fig. 6.21 FIB section B, Region of Interest 2 (see Fig. 6.15e). **a** Interstitial aluminum alloy between the QCs has a mottled contrast attributable to local strain and twinning. **b** [110] zone of the *fcc* Al-alloy. **c** [211] zone of the *fcc* Al-alloy. The double spots indicate twinning by reticular pseudo-merohedry. Modified after Oppenheim et al. (2017b)

ratio of the major metals $Al_{76}Ni_{20}Fe_4$ (Table 6.2) is quite similar to the natural case, $Al_{71}Ni_{24}Fe_5$. The same resemblance is evident if we take into account the layer spacing perpendicular to the 10-fold symmetry axis: 4.2 versus 4.204(2) Å for synthetic and natural d-QC, respectively. Another important similarity is the absence of phason strain. Indeed, any weak diffuse reflections, characteristic of imperfectly ordered atoms in the quasi-periodic 10-fold layers, was observed in the electron diffraction pattern collected down the 10-fold axis (Fig. 6.16c). Such perfection is astonishing given the dynamic synthesis process and the fine intergrowth with other crystalline phases, but it is exactly the same consideration made for both the shock-produced icosahedral quasicrystals (see Sect. 6.2) and for the natural occurrences. In the latter case, it was hypothesized that the low phason strain observed might be a side-effect of the original growth process or it might reflect annealing over the very long age of the specimen (i.e., billions of years). In the experimental products we have seen and described in this Chapter, however, we had nanoseconds annealing times thus sustaining the notion that shock synthesis grows essentially strain-free highly perfect quasicrystals from the outset.

Tsai et al. (1989) first discovered decagonal $Al_{70}Ni_{15}Fe_{15}$ quasicrystals in the Al–Ni–Fe system. By means of a convergent-beam electron diffraction (CBED) study, Saito et al. (1992) found the so-called G-M lines (Gjønnes-Moodie lines) in odd-order reflections along the c^* direction, showing that these quasicrystals belong to the non-centrosymmetric space group $P\overline{1}0m2$.[1] Indeed, the presence of glide planes or screw axes in a crystal structure generally introduces dark bars (for kinematically forbidden reflections) known as G-M lines. The findings by Saito et al. (1992) were confirmed in a subsequent TEM study by Tsuda et al. (1993). Using CBED and high-resolution microscopy, Tanaka et al. (1993), revealed a transition from $P\overline{1}0m2$ to the centrosymmetric $P10_5/mmc$ as a function of the Ni/Fe ratio. In detail, along the join $Al_{70}Ni_xFe_{30-x}$ in the range $10 < x < 20$, these authors showed that alloys with $10 < x < 17.5$ crystallize with the $P\overline{1}0m2$ structure, whereas alloys with $17.5 < x < 20$ give rise to the $P10_5/mmc$ structure.

For the shock-produced decagonal quasicrystal, all the TEM-SAEDs we collected that have the c^* direction display clear odd-order reflections, indicating the absence of the c-glide along the 10-fold axis. This likely indicates that the 10_5 screw axis is absent and the d-QC studied here exhibits the non-centrosymmetric $P\overline{1}0m2$ structure. Another possibility is that it exhibits a different centrosymmetric space group such as $P10_5/mmm$.

It is noteworthy that the composition of the d-QC reported here is close to that of the thermodynamically stable decagonal phase in the Al–Ni–Fe system, $Al_{71}Ni_{24}Fe_5$ (Lemmerz et al. 1994), which is known to exhibit the centrosymmetric $P10_5/mmc$ structure (Saitoh et al. 2001). The same results were found for the natural analogue of $Al_{71}Ni_{24}Fe_5$, the mineral decagonite, by single-crystal X-ray diffraction (Bindi et al. 2015b). Although it is possible that our shock-recovered specimen occupies a metastable or preserved high-pressure structure, we speculate that the discrepancy in the space group could be due to the presence of other minor elements in the chemical formula. Ordering of such minor elements could definitely influence the atomic arrangements sufficiently to make the difference among these space groups.

6.9   What if We Shock an Already Formed Quasicrystal?

After the successful synthesis of icosahedral and decagonal quasicrystals by shock, we designed a new experiment having an already formed quasicrystal as target. This was considered to be important in order to shed light on the formation of i-phase *II* (Bindi et al. 2016). As starting material of the shock experiment, we used the Sigma-Aldrich 757934-10G icosahedral AlCuFe quasicrystal powder. The grains in the powder are mostly spherical with sizes from 1 to 20 μm. XRD shows predominance of icosahedral quasicrystals plus minor Al–Fe alloys in the powder. Electron microprobe analyses indicate that the composition of the quasicrystal is $Al_{62-64}Cu_{23-25}Fe_{12-14}$.

---

[1]The symbol looks like that of a 3D space-group type but it is actually a superspace group. The superspace approach (de Wolff 1974) allows us to describe a quasicrystal as a periodic object in the real world by considering it as a three-dimensional intersection of this periodic object in a higher dimension. For a more extensive review about how to determine space groups or how to embed them in superspace please refer to Takakura (2010).

The sample chamber in the SS304 recovery assembly was machined to 0.5" diameter to fit the KBr pressure pin. About 0.09 g of the starting QC powder was pressed to 0.5 GPa in the chamber. The bulk thickness of the compressed powder was 0.45–0.50 mm. A SS304 pin and vacuum epoxy were used to close up sample chamber after pressing the QC powder. The impact velocity of the tantalum flyer for shot S1250 was 0.93 km/s, resulting in 23.8 GPa shock pressure in the SS304 chamber. Assuming the quasicrystal sample is thin enough to equilibrate with the chamber through reverberation, the peak shock pressure in the sample would be the same 23.8 GPa.

The front (driver) of the recovery chamber broke apart at the sample-driver interface during the shock experiment (Fig. 6.22). Most of the shocked quasicrystal was attached to the surface of the driver and chamber pieces and was nevertheless recovered.

The recovered sample consisted of semi-spherical (circular in 2D projection) grains of 10–20 μm in sizes (Fig. 6.23). This granular microstructure is inferred to result from the starting quasicrystal powder, although the interstitial pore space was closed up by shock compression. Within the grains, two distinctly different microstructures, a decomposition and a homogeneous microstructure, are observed.

First, the quasicrystal grain decomposed into two phases of different BSE contrast (Figs. 6.23 and 6.24). The phase with high BSE contrast is enriched in copper. EDS analyses with 10–15 kV accelerating voltage provide a composition of $Al_{58}Fe_{10}Cu_{32}$. Because the bright grains are at most sub-micron in size (Fig. 6.24), interference from the surrounding phases is not avoidable with SEM-EDS. Electron back-scattered patterns (EBSPs) of this phase are indexed as a face-centered cubic structure ($Im\bar{3}m$), with a common mean angle deviation (MAD) number of 0.6°.

Given the observed structure and composition, this Al-alloy could be a Cu-rich steinhardtite. However, since the $Im\bar{3}m$ and $Pm\bar{3}m$ space group types show very similar features using EBSD and the composition measurement is likely interfered (see above), TEM was needed to confirm the occurrence of steinhardtite. On the other hand, the phase of dark BSE contrast in the decomposed grains show A5, A3 and A2 rotation axis in its EBSPs, indicating icosahedral symmetry (Fig. 6.24). EDS analyses provide a composition of $Al_{62}Cu_{25}Fe_{13}$. This composition is almost identical to the

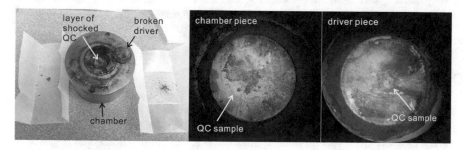

Fig. 6.22 Images of the broken target assembly. The center circle of the chamber and driver pieces is 0.5" in diameter

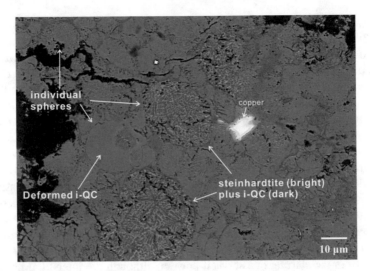

Fig. 6.23 BSE images of the shocked icosahedral quasicrystal. The grains with mottled microstructure are decomposed to two phases

EMPA result of the starting material ($Al_{62-64}Cu_{23-25}Fe_{12-14}$), resulting in difficulty in explaining the Cu-rich alloy by decomposition. Nevertheless, bulk composition of the decomposed grain is consistently $Al_{61}Cu_{26-27}Fe_{12-13}$ by EDS. This result shows higher copper content than the EMPA of the starting material. Even assuming that difference is a systematic error from EDS, the shock-induced i-QC phase is still slightly more Cu-deficient than the bulk grain and the starting material.

Portions of some grains show a smooth homogenous BSE contrast (Fig. 6.25). EBSD indicates that the smooth region consists of pure icosahedral quasicrystal. However, the i-QC occurs in sub-micron domains of random crystallographic orientations. Their band quality of diffraction pattern is generally worse than that of the i-QC from decomposition (Fig. 6.25). This feature might result from shock deformation. The composition of the smooth regions is consistently $Al_{61}Cu_{27}Fe_{12}$, which is the same as the bulk composition of the decomposed grains by EDS analysis. It is likely that the shock-metamorphosed grains have the same bulk composition as the starting material but our EDS analysis shows a systematic error.

6.10 Producing i-Phase *II* by Shock

The composition of the first reported natural quasicrystal (icosahedrite, $Al_{65}Cu_{23}Fe_{12}$) matches with its stability field in the Al–Cu–Fe system (Tsai et al. 1987). However, the second natural discovered icosahedral quasicrystal (i-phase *II*, $Al_{61}Cu_{32}Fe_7$) (Bindi et al. 2016) was found to be outside the stability field at ambient pressure and has not been synthesized before. It is also compositionally different

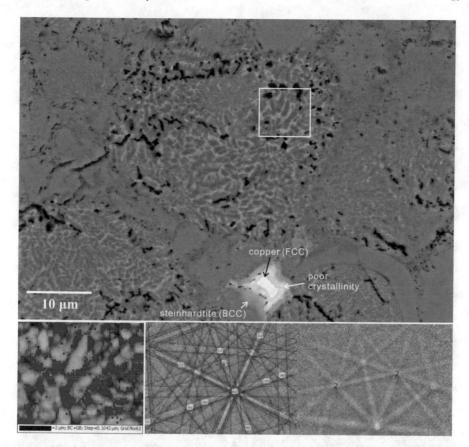

Fig. 6.24 BSE image and EBSD map of decomposed grain. Top: BSE image in variable pressure mode. The box outlines the area for EBSD analysis. Bottom-left: EBSD band contrast and phase map. High brightness indicates high Kikuchi band quality. Areas indexed as *bcc* structure correspond to the bright-contrast phase in the BSE image and are marked in blue. The uncolored areas show icosahedral EBSPs, corresponding to the phase of dark BSE contrast. Bottom-center: indexed EBSP of the *bcc* alloy. Bottom-right: icosahedral EBSP including A5, A3 and A2 axes

from the icosahedral quasicrystals reported in the shock-recovery studies described above. For this reason, we decided to carry out two new shock-recovery experiments using an Al–Cu–W graded density disk (GDI), originally manufactured as a ramp-generating impactor but here used as target, to sample a wide range of Al/Cu ratios and traverse the full stability field of icosahedral Al–Cu–Fe quasicrystals in compositional space (Figs. 6.26 and 6.27). To our surprise, we were able to easily synthesize i-phase *II* (Figs. 6.28 and 6.29) in these experiments.

The composition and the structure (icosahedral) of these shock-produced samples are nearly identical to those described for the natural analog from the Khatyrka meteorite (Fig. 6.30).

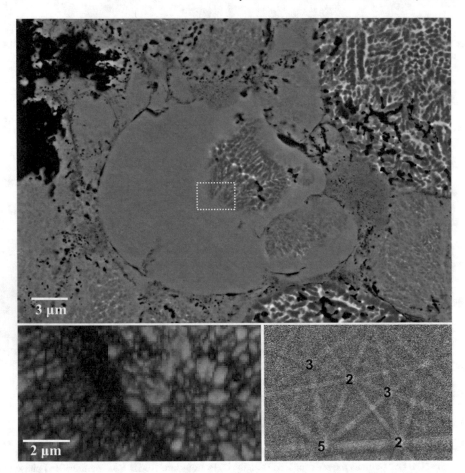

Fig. 6.25 Top: BSE image of a smooth i-QC grain containing a decomposed region. The boxed area is mapped with EBSD. Bottom-right: EBSD band contrast map. The dark area is the boundary between the smooth and decomposed regions. The smooth region consists of small domains and generally worse pattern quality than the decomposed region. Bottom-left: representative i-QC EBSP from the smooth region

It is noteworthy that, although the observed intermetallic phases are Al-rich, the i-phase *II* bearing region in the run product starts with relatively high Cu content. In the previous shock experiments that used Al_5Cu as starting material, the shock-induced i-phase has higher Al content than natural samples (Asimow et al. 2016; Oppenheim et al. 2017a, b). Strong shear flow is very likely required to form the icosahedral quasicrystal, particularly i-phase *II*, during shock. In most of our shock experiments, reactions along the less deformed region forms only intermetallic alloys.

The new experiments suggest that the optimal $Al_{65}Cu_{23}Fe_{12}$ icosahedral quasicrystal composition is not the most preferred in this (15–30 GPa) shock condition,

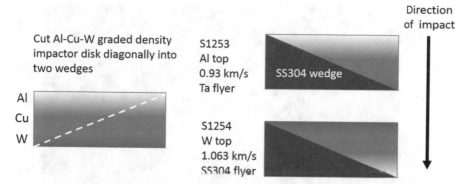

Cut Al-Cu-W graded density impactor disk diagonally into two wedges

S1253
Al top
0.93 km/s
Ta flyer

S1254
W top
1.063 km/s
SS304 flyer

SS304 wedge

Direction of impact

Al
Cu
W

Fig. 6.26 Schematic of the impactors design in the new shock-recovery experiments using an Al–Cu–W graded density disk

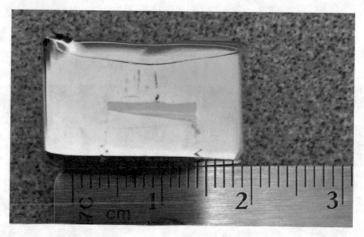

Fig. 6.27 Experimental chamber of the new experiments using the Al–Cu–W graded density disk

although it is thermodynamically stable under lower pressures. This seems to suggest that multiple stages of impacts of the Khatyrka meteorite could account for the coexisting i-phase *I* (icosahedrite) and i-phase *II* contained in it.

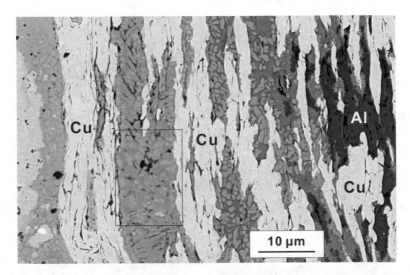

Fig. 6.28 Reaction among steel, Cu and Al in the Cu-rich portion of the graded density impactors with strong deformation. Boxed area is enlarged in next BSE images

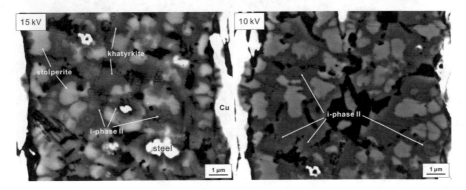

Fig. 6.29 Left: Petal-like icosahedral quasicrystal (i-phase *II*), entrained in intermetallic phases, including stolperite (β, CuAl, *Pm$\bar{3}$m*) and khatyrkite (θ, CuAl$_2$, *I4/mcm*). Right: Lower accelerating voltage (10 kV) BSE image of i-phase *II*. The contrast is weak because of smaller electron excitation volume

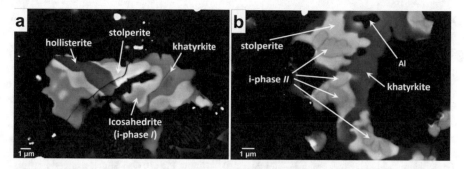

Fig. 6.30 Natural occurrences of icosahedrite and i-phase *II* in the Khatyrka meteorite in association with hollisterite (λ, Al_3Fe, *C2/m*), stolperite and khatyrkite (Bindi et al. 2016)

References

Asimow PD, Lin C, Bindi L, Ma C, Tschauner O, Hollister LS, Steinhardt PJ (2016) Shock synthesis of quasicrystals with implications for their origin in asteroid collisions. Proc Nat Acad Sci USA 113:7077–7081

Bindi L, Lin C, Ma C, Steinhardt PJ (2016) Collisions in outer space produced an icosahedral phase in the Khatyrka meteorite never observed previously in the laboratory. Sci Rep 6:38117

Bindi L, Steinhardt PJ, Yao N, Lu PJ (2009) Natural quasicrystals. Science 324:1306–1309

Bindi L, Steinhardt PJ, Yao N, Lu PJ (2011) Icosahedrite, $Al_{63}Cu_{24}Fe_{13}$, the first natural quasicrystal. Am Min 96:928–931

Bindi L, Eiler J, Guan Y, Hollister LS, MacPherson GJ, Steinhardt PJ, Yao N (2012) Evidence for the extra-terrestrial origin of a natural quasicrystal. Proc Nat Acad Sci USA 109:1396–1401

Bindi L, Yao N, Lin C, Hollister LS, Andronicos CL, Distler VV, Eddy MP, Kostin A, Kryachko V, MacPherson GJ, Steinhardt WM, Yudovskaya M, Steinhardt PJ (2015a) Natural quasicrystal with decagonal symmetry. Sci Rep 5:9111

Bindi L, Yao N, Lin C, Hollister LS, Andronicos CL, Distler VV, Eddy MP, Kostin A, Kryachko V, MacPherson GJ, Steinhardt WM, Yudovskaya M, Steinhardt PJ (2015b) Decagonite, $Al_{71}Ni_{24}Fe_5$, a quasicrystal with decagonal symmetry from the Khatyrka CV3 carbonaceous chondrite. Am Min 100:2340–2343

Bindi L, Yao N, Lin C, Hollister LS, Poirier GR, Andronicos CL, MacPherson GJ, Distler VV, Eddy MP, Kostin A, Kryachko V, Steinhardt WM, Yudovskaya M (2014) Steinhardtite, a new body-centered-cubic allotropic form of aluminum from the Khatyrka CV3 carbonaceous chondrite. Am Min 99:2433–2436

Cahn J, Shechtman D, Gratias D (1986) Indexing of icosahedral quasiperiodic crystals. J Mat Res 1:13–26

Chumak I, Richter KW, Ipser H (2007) The Fe–Ni–Al phase diagram in the Al-rich (> 50at.% Al) corner. Intermetallics 15:1416–1424

de Wolff PM (1974) The pseudo-symmetry of modulated crystal structures. Acta Crystallogr A 30:777–785

Hollister LS, Bindi L, Yao N, Poirier GR, Andronicos CL, MacPherson GJ, Lin C, Distler VV, Eddy MP, Kostin A, Kryachko V, Steinhardt WM, Yudovskaya M, Eiler JM, Guan Y, Clarke JJ, Steinhardt PJ (2014) Impact-induced shock and the formation of natural quasicrystals in the early Solar System. Nat Comm 5:3040

Kenkmann T, Hornemann U, Stöffler D (2000) Experimental generation of shock-induced pseudotachylites along lithological interfaces. Met Plan Sci 35:1275–1290

Lemmerz U, Grushko B, Freiburg C, Jansen M (1994) Study of decagonal quasicrystalline phase formation in the Al–Ni–Fe alloy system. Phil Mag Lett 69:141–146

Levine D, Lubensky TC, Ostlund S, Ramaswamy A, Steinhardt PJ (1985) Elasticity and defects in pentagonal and icosahedral quasicrystals. Phys Rev Lett 54:1520–1523

Lu PJ, Deffeyes K, Steinhardt PJ, Yao N (2001) Identifying and indexing icosahedral quasicrystals from powder diffraction patterns. Phys Rev Lett 87:275507

Lubensky TC, Socolar JES, Steinhardt PJ, Bancel PA, Heiney PA (1986) Distortion and peak broadening in quasicrystal diffraction patterns. Phys Rev Lett 57:1440–1443

MacPherson GJ, Andronicos CL, Bindi L, Distler VV, Eddy MP, Eiler JM, Guan Y, Hollister LS, Kostin A, Kryachko V, Steinhardt WM, Yudovskaya M, Steinhardt PJ (2013) Khatyrka, a new CV3 find from the Koryak Mountains, Eastern Russia. Met Plan Sci 48:1499–1514

Malavergne V, Guyot F, Benzerara K, Martinez I (2001) Description of new shock-induced phases in the Shergotty, Zagami, Nakhla and Chassigny meteorites. Met Plan Sci 36:1297–1305

Oppenheim J, Ma C, Hu J, Bindi L, Steinhardt PJ, Asimow PD (2017a) Shock synthesis of five-component icosahedral quasicrystals. Sci Rep 7:15629

Oppenheim J, Ma C, Hu J, Bindi L, Steinhardt PJ, Asimow PD (2017b) Shock synthesis of decagonal quasicrystals. Sci Rep 7:15628

Potter DK, Ahrens TJ (1994) Shock induced formation of $MgAl_2O_4$ spinel from oxides. Geophys Res Lett 21:721–724

Quiquandon M, Quivy A, Devaud J, Faudot F, Lefebvre S, Bessière M, Calvayrac Y (1996) Quasicrystal and approximant structures in the Al–Cu–Fe system. J Phys Condens Matter 8:2487

Saito M, Tanaka M, Tsai A-P, Inoue A, Masumoto T (1992) Space group determination of decagonal quasicrystals of an $Al_{70}Ni_{15}Fe_{15}$ alloy using convergent-beam electron diffraction. Jap J Appl Phys 31:L109

Saitoh K, Tanaka M, Tsai A-P (2001) Structural study of an $Al_{73}Ni_{22}Fe_5$ decagonal quasicrystal by high-angle annular dark-field scanning transmission electron microscopy. J Electr Micr 50:197–203

Sharp TG, DeCarli PS (2006) Meteorites and the early solar system II. In: Lauretta DS, McSween HY Jr (eds) University of Arizona Press, pp 653–677

Stagno V, Bindi L, Park C, Tkachev S, Prakapenka VB, Mao H-K, Hemley RJ, Steinhardt PJ, Fei Y (2015) Quasicrystals at extreme conditions: the role of pressure in stabilizing icosahedral $Al_{63}Cu_{24}Fe_{13}$ at high temperature. Am Min 100:2412–2418

Steurer W, Deloudi S (2009) Crystallography of quasicrystals. Concepts, methods and structures. Springer Series in Materials Science, vol 126. Springer, Heidelberg

Stöffler D, Bischoff A, Buchwald V, Rubin A (1988) Meteorites and the early solar system. In: Kerridge JF, Matthews MS (eds) University of Arizona Press, pp 165–202

Takakura H (2010) Decagonal quasicrystals: higher dimensional description and structure determination. International school on aperiodic crystals 26 Sept–2 Oct 2010, La Valérane Carqueiranne, France

Tanaka M, Tsuda K, Terauchi M, Fujiwara A, Tsai A-P, Inoue A, Masumoto T (1993) Electron diffraction and electron microscope study on decagonal quasicrystals on Al–Ni–Fe alloys. J Non-Crystall Solids 153–154:98–102

Tomeoka K, Yamahana Y, Sekine T (1999) Experimental shock metamorphism of the Murchison CM carbonaceous chondrite. Geochim Cosmochim Acta 63:3683–3703

Tsai A-P, Inoue A, Masumoto T (1987) A stable quasicrystal in Al–Cu–Fe system. Jap J Appl Phys 26:L1505

Tsai A-P, Inoue A, Masumoto (1989) New decagonal Al–Ni–Fe and Al–Ni–Co alloys prepared by liquid quenching. Mat Trans JIM 30:150–154

Tsuda K, Saito M, Terauchi M, Tanaka M, Tsai A-P, Inoue A, Masumoto T (1993) Electron microscope study of decagonal quasicrystals of $Al_{70}Ni_{15}Fe_{15}$. Jap J Appl Phys 32:129

Willis M, Ahrens T, Bertani L, Nash C (2006) Bugbuster—survivability of living bacteria upon shock compression. Earth Plan Sci Lett 247:185–196

Yamamoto A, Tsubakino H (1997) Al_9Ni_2 precipitates formed in an Al–Ni dilute alloy. Scripta Mater 37:1721–1725

Zhang L, Du Y (2007) Thermodynamic description of the Al–Fe–Ni system over the whole composition and temperature ranges: modeling coupled with key experiment. Calphad 31:529–540

Zhang L, Lück R (2003) Phase diagram of the Al–Cu–Fe quasicrystal-forming alloy system: I. Liquidus surface and phase equilibria with liquid. Z Metallkd 94:91–97

Chapter 7
Why Do Quasicrystals Grow in Asteroidal Collisions?

As we have already seen, to shed light on the mechanisms leading to the formation of quasicrystals during hypervelocity asteroidal collisions in outer space, we designed several shock experiments (Chap. 6). But what about the heating mechanisms during the shock? To try to answer to this question, I will focus on shock-produced Al–Cu–Fe quasicrystals.

Taking into account the evaluation of shock temperatures for dense starting materials given by Ahrens (1987), we can say with confidence that the T was always <350 °C in the shock experiments described in Chap. 6, and that post-shock release temperatures were even lower. Noteworthy, $T < 350$ °C are well below (at both shock and ambient pressures) the melting points of all the materials used in the experiments. However, we have observed clear features (e.g., metal blebs, amorphous quench phases) of local melting in the various parts of the recovered sample chambers. Such features also include the nucleation and growth of quasicrystal itself. This does mean that other local heating mechanisms acted to melt the starting materials, which was not a thermite reaction. Let's try to consider porosity and shear heating as main mechanisms that could have affected the experiments and then try to quantify whether either or both could explain the formation of the melted regions we have observed.

It is important to state that during the preparation of the sample (machining and polishing) we could have created some cavities and voids along the boundary between $CuAl_5$ and the SS304. Under shock, such cavities would collapse thus producing additional heat in the vicinity of the collapsed voids. Such a situation can be modeled using a standard continuum porosity model (Ahrens et al. 1990) by considering the boundary as a thin layer of porous $CuAl_5$ sandwiched between the bulk, fully dense $CuAl_5$ and the SS304. It can be easily demonstrated that the presence of only 0.5% porosity is able to raise $T_{shock} > 550$ °C, which is the melting temperature (at ambient pressure) of $CuAl_2$–Al mixtures. At 20 GPa, the amount of porosity to melt pure Al or Cu is 2.5%. As a consequence, to reproduce the features we observe in the shock-produced sample, it is sufficient to hypothesize the presence of just few 0.25 μm-sized

© The Author(s), under exclusive license to Springer Nature Switzerland AG 2020
L. Bindi, *Natural Quasicrystals*, SpringerBriefs in Crystallography,
https://doi.org/10.1007/978-3-030-45677-1_7

voids along the capsule wall, which can easily induce melting of a 10 μm thick layer of CuAl$_5$.

Now let's consider the mostly fine-grained, mixed layer (20 μm thick) reported in Fig. 6.7c. Taking into account the difference in the impedance properties of SS304 and CuAl$_5$, the propagation of a shockwave parallel to this two-materials boundary produces a different velocity across the interface. Such a difference produces shear, which is also due to the different shock velocity in the two materials and the different particle velocity. Although it is very hard to quantify a viscous shear-heating model because of the lack of information on the viscosity within the developing mixed layer, we can try to model it as a dry frictional sliding problem using a solid cylinder of CuAl$_5$ sliding through a hollow cylinder of SS304, and verify if the coefficient of friction needed to melt the materials is conceivable.

The rate of heat dissipation into the CuAl$_5$ layer, q, caused by frictional sliding, per unit area is

$$q = \alpha \mu \Delta P \Delta U \qquad (7.1)$$

where α is the fraction of the heat generation partitioned to the CuAl$_5$ side of the boundary (assumed to be 0.5, given the characteristics of the material), μ is the coefficient of kinetic friction, ΔP is the normal stress across the boundary (considered equivalent to the difference between first shock pressures in SS304 and CuAl$_5$, 8.8 GPa), and ΔU is the velocity difference across the boundary (difference in first-shock particle velocity in the two materials, 184 ms^{-1}). The shock-time-pulse can be quantified to be approximately $t = 0.7$ μs. By assuming that: (i) the total heat dissipated during the sliding is deposited into a 20 μm CuAl$_5$ layer (e.g., Fig. 6.7c), (ii) an (initial) density ~3570 kg m^{-3}, and (iii) a mass per unit area of heated material of $m = 0.07$ kg m^{-2}, then the total energy dissipated during the sliding event can be written as:

$$\Delta E = \frac{qt}{m} = 8\mu \text{ MJ.} \qquad (7.2)$$

The temperature increase, assuming a constant specific heat $c_v = $ ~700 J kg^{-1} K^{-1} (estimated from the Kopp-Neumann mixing law), is

$$\Delta T = \frac{\Delta E}{C_v} = 10^4 \mu \text{ K.} \qquad (7.3)$$

Taking into account a $T_{\text{shock}} < 350$ °C (see above) and a post-shock temperature > 550 °C, a μ value >0.02 is required for melting upon release. Much higher values of μ (>0.14) are instead needed to obtain the melting of either pure Al or Cu at 20 GPa (both ~1700 °C). The obtained values seem reasonable by considering the typical coefficients of friction. This seems to suggest that sliding friction on the side-walls of the chamber could be a reasonable mechanism to explain the observed melted textures.

References

Ahrens TJ (1987) Shock wave techniques for geophysics and planetary physics. In: Sammis CG, Henyey TL (eds) Methods of experimental physics. Academic Press, New York

Ahrens TJ, Tan H, Bass JD (1990) Analysis of shock temperature data for iron. Int J High Press Res 2:145–157

Chapter 8
On the Stability of Quinary Quasicrystals

We have seen that the composition of the quasicrystals obtained in the various shock experiments is heretofore unreported. For this reason, it is not straightforward to understand if we are dealing with a stable or metastable quasicrystal at ambient conditions. If we consider the Al–Cu–Fe abundances alone, the new shock-produced i-phases are clearly outside the stability field of the quasicrystalline phase reported in the Al–Cu–Fe ternary system. For this reason, it is useful to consider if the presence of additional elements might affect their stability. The concept of stability of quasicrystals is usually studied taking into account the following empirical methods: Hume-Rothery rules and the cluster line approach. Such methods are well developed for ternary quasicrystals (e.g., Mizutani 2010; Dong et al. 2007, respectively). Here an attempt to extend the two methods to five components is presented.

8.1 Hume-Rothery Phase Criterion

The Hume-Rothery rules (Mizutani 2010) predict the stabilization of the various alloys on the basis of electrochemical, size, and valence electron factors. With "Hume-Rothery phase" we refer to a phase stabilized by its valence electron concentration—its electron density does contain a pseudo-gap at the Fermi level due to interference with the Jones zone. This is usually done using the criterion of constant (or nearly constant) values of electrons per atom (e/a). Given the Fermi energy E_f and the density of states $N(E)$,

$$(e/a) \equiv \int_0^{E_f} N(E)\mathrm{d}E \qquad (8.1)$$

Belin-Ferré (2010). The (e/a) value of an alloy can be obtained from the atomic fractions n_i and the valence V_i of each element in the alloy using the rule

© The Author(s), under exclusive license to Springer Nature Switzerland AG 2020
L. Bindi, *Natural Quasicrystals*, SpringerBriefs in Crystallography,
https://doi.org/10.1007/978-3-030-45677-1_8

Table 8.1 Valence and atomic radii to be used for quasicrystal stability (Dong et al. 2004)

	Al	Cu	Ni	Fe	Cr
Mott and Jones valence	+3	+1	0	−2	−4
Raynor valence	+3	+1	−0.61	−2.66	−4.66
Atomic radius (pm)	143	128	124	126	128

$$(e/a) \approx \sum_i V_i n_i, \tag{8.2}$$

but the correct and real valences of transition metals are dependent on the composition (see Table 8.1 for most common used values), which brings in some ambiguity into this analysis. Mott and Jones valences are assumed here for the follow-up discussion.

To build up compositional criteria for quasicrystal stability, the following closure constraint is assumed:

$$\sum_i n_i = 1 \tag{8.3}$$

Then, we have to take into consideration that quasicrystalline phases in the Al–Cu–Fe system have been experimentally shown to exhibit $(e/a) = 1.862$ (Mizutani 2010) and thus satisfying

$$\sum_i V_i n_i = 1.862. \tag{8.4}$$

For a ternary system, the solution of Eqs. (8.3) and (8.4) is an (e/a)-constant line across the ternary (with one degree of freedom), which has been helpful to envisage the right location of quasicrystals in ternary systems. For a quaternary system, the solution is represented by a plane, which can be reduced using one additional constraint. Indeed, the Hume-Rothery rules provide a further restraint on the average atomic size R_a (Dong et al. 2004), between 137 and 140 pm, for icosahedral quasicrystals. Thus, if R_i is the radius of each atom composing the alloy (Table 8.1), the additional constraint is

$$137 \text{ pm} < R_a = \sum_i R n_i < 140 \text{ pm}, \tag{8.5}$$

which decreases the solution set for a quaternary system to a narrow band and makes such an approach again useful to predict where a quasicrystal is stabilized (Dong et al. 2007).

Assuming that the same stability criteria can be applied also for the Al–Cu–Fe–Cr–Ni quinary system, the region for potential stable quasicrystals can be individuated using the same constraints. The solution of Eqs. (8.3) and (8.4) in 5-D space,

$$n_{Fe} + 1.4n_{Cr} + 0.6n_{Ni} + 0.4n_{Cu} = 0.2276, \qquad (8.6)$$

is now a volume whose projection into the tetrahedral 3-D volume Al–Cu–Fe–(Cr + Ni) (bottom in Fig. 8.1) is delimited by the surfaces depicting the Al–Cu–Fe–Cr and Al–Cu–Fe–Ni stability fields. Again, taking into account the size constraint 138 $< R_a <$ 139 pm (to avoid negative solutions that occur near 137 or 140 pm), it lowers the solution set to a surface, represented in the top part of Fig. 8.1b.

8.2 Cluster Line Approach

The concept of stability of ternary quasicrystals has been also studied using the cluster-line approach (Dong et al. 2007), which consider a 'cluster' as a stable binary unit, somewhat comparable to a unit cell in periodic crystals. The cluster-and-glue model divides the atoms into clusters and glue atoms dispersed between them. In a ternary diagram, a cluster line is drawn from a known binary cluster and a third component. If in the system two binary clusters from different pairs of components are present, the lines will cross and the composition of the junction usually represents that of a stable quasicrystal. Some of the binary clusters of Al–Cu–Fe–Cr–Ni are shown in Table 8.2. If we now move to a five-dimensional system, there will be the appearance of cluster hyperplanes rather than cluster lines. However, the stability is defined by the junction of four hyperplanes, which still represents a point in the normalized compositional space.

The solution of Eq. (8.3) and of all the four primary-cluster-hyperplanes intersection is the composition $Al_{54}Cu_9Fe_{10}Cr_{10}Ni_{16}$ (projected to the Al–Cu–Fe ternary as red square symbol in the bottom part of Fig. 8.1). Because of the small amount of Cr and Ni in the studied quasicrystals, it is likely that the stability is governed by the Al–Cu and Al–Fe clusters. Taking into account the: (i) intersection of the primary cluster lines for Al–Cu and Al–Fe pairs, (ii) the solution of Eq. (8.6) and (iii) the average size constraint 137.4 pm $< R_a <$ 138.4 pm, the stable composition is in the range $Al_{67-71}Cu_{11-12}Fe_{12-13}Cr_{0-3}Ni_{0-9}$.

As shown in Fig. 8.1, icosahedrite (Bindi et al. 2009, 2011) plots along the $(e/a) =$ 1.862 line near the intersection of the Al_8Cu_3 and $Al_{10.7}Fe_2$ clusters. The geometry of the Al_8Cu_3 cluster is a octahedral antiprism and that of $Al_{10.7}Fe_2$ is an icosahedron. The combination of these two clusters results in a Mackay icosahedron. The five-component icosahedral AlCuFe-quasicrystals synthesized in the shock experiments (Asimow et al. 2016; Oppenheim et al. 2017) all plot near the intersection of $Al_{11.1}Cu_{1.9}$ and $Al_{10.7}Fe_2$. Both of these clusters are icosahedra from a geometric point of view, suggesting that these quasicrystals consist of Bergman clusters (Dong et al. 2007).

It is actually uncertain to say if Hume-Rothery constraints or cluster lines can really be helpful to give insights on the stabilization of five-component icosahedral quasicrystals, also because the (e/a) values determined from our microprobe chemical data vary from 1.84 to 2.07.

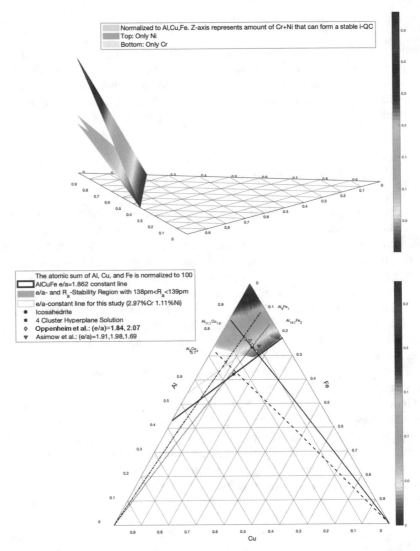

Fig. 8.1 (top) Projection of the volume of (*e/a*)-constant stability criteria from five-component Al–Cu–Fe–Cr–Ni space to the ternary Al–Cu–Fe diagram (see text for explanation). The planes represent the amount of Ni, Cr, or Ni + Cr in a 3:1 ratio to have a stable quasicrystal. The different colors indicate the atomic fraction of Cr + Ni. (bottom) The position of real and predicted quasicrystals in the Al–Cu–Fe–Cr–Ni system, projected onto the Al–Cu–Fe ternary system. Symbols: (*e/a*)-constant for ternary Al–Cu–Fe (thick red), (*e/a*)-constant for the atomic fractions of Cr and Ni (thin yellow), various cluster lines, Al₆₃Cu₂₄Fe₁₃ icosahedrite (red circle), new compositions from Oppenheim et al. (2017) (diamonds), new compositions from Asimow et al. (2016) (downward triangles), the 4-cluster hyperplane solution (red square), and the estimated stability region for five-component quasicrystals given by (*e/a*) and size constraints (contoured surface, colors indicate atomic fraction of Cr + Ni). Modified after Oppenheim et al. (2017)

Table 8.2 Stable clusters in Al–TM binaries (Chen et al. 2011)

Binary	Primary cluster	Secondary cluster
Al–Cu	$Al_{11.1}Cu_{1.9}$	Al_8Cu_3
Al–Fe	$Al_{10.7}Fe_2$	Al_9Fe_4
Al–Cr	$Al_{11}Cr_2$	$Al_{12}Cr_1$
Al–Ni	$Al_{10}Ni_3$	Al_9Ni_3

Our shock synthesis experiments produced for the first time quasicrystals in a new range of composition. To study in more detail the limits of their stability field, however, calls for new careful conventional metallurgical synthesis of larger, more uniform materials.

References

Asimow PD, Lin C, Bindi L, Ma C, Tschauner O, Hollister LS, Steinhardt PJ (2016) Shock synthesis of quasicrystals with implications for their origin in asteroid collisions. Proc Nat Acad Sci USA 113:7077–7081

Belin-Ferré E (2010) Properties and applications of complex intermetallics. World Scientific

Bindi L, Steinhardt PJ, Yao N, Lu PJ (2009) Natural quasicrystals. Science 324:1306–1309

Bindi L, Steinhardt PJ, Yao N, Lu PJ (2011) Icosahedrite, $Al_{63}Cu_{24}Fe_{13}$, the first natural quasicrystal. Am Min 96:928–931

Chen H, Qiang J, Wang Q, Wang Y, Dong C (2011) A cluster-resonance criterion for Al-TM quasicrystal compositions. Israel J Chem 51.1226–1234

Dong C, Wang Q, Qiang J, Wang Y, Chen W, Shek CH (2004) Composition rules from electron concentration and atomic size factors in Zr-Al-Cu-Ni bulk metallic glasses. Mater Trans 45:1177–1179

Dong C, Wang Q, Qiang J, Wang Y, Jiang N, Han G, Li Y, Wu J, Xia J (2007) From clusters to phase diagrams: composition rules of quasicrystals and bulk metallic glasses. J Phys D Appl Phys 40:R273

Mizutani U (2010) Hume-Rothery rules for structurally complex alloy phases. Taylor & Francis. ISBN 978-1-4200-9058-1

Oppenheim J, Ma C, Hu J, Bindi L, Steinhardt PJ, Asimow PD (2017) Shock synthesis of five-component icosahedral quasicrystals. Sci Rep 7:15629

Chapter 9
Are Quasicrystals Really so Rare in the Universe?

The discovery of icosahedrite, the first natural quasicrystal, and the subsequent discovery of two other types of quasicrystals challenged the conventional wisdom. Icosahedrite was found to be formed naturally in CV3 chondrites that comprised the primordial material of our Solar System. So, it is not impossible to form it outside the laboratory. Since it has only been reported in one CV3 chondrite to date, one cannot draw precise quantitative conclusions. However, since few CV3 chondrites have been studied with the same detail as the Khatyrka meteorite, one might suppose that this was a one-in-a-thousand event; or maybe one in a million. Even so, there is a lot of similar pre-planetary material around our Solar System, in stellar systems in our galaxy, and in galaxies throughout the universe. The number of different minerals in these pre-planetary materials may number 50 or so. Compare that to most of the minerals we know on Earth which require the formation of complete planets and the oxygenation of their atmospheres. They are likely much, much rarer. So, by this measure, quasicrystals may be much more common in the universe as a whole than scientists originally expected.

Furthermore, natural quasicrystals have been reported only in the Khatyrka meteorite so far, but Al-bearing alloys have been also found in the shocked Suizhou L6 chondrite (Xie and Chen 2016), in the Zhamanshin impact structure (Gornostaeva et al. 2018), and in the carbonaceous, diamond-bearing stone "Hypatia" (Belyanin et al. 2018). In addition, a mineralogical assemblage nearly identical to that observed in Khatyrka has been recently described in a cosmic spherule recovered in the Nubian desert, Sudan (Suttle et al. 2019). So, it is very likely that other quasicrystals will be discovered soon, now that the first discovery (Bindi et al. 2009) is settled and well accepted by the scientific community.

A crucial point in the process of discovery will be the attitude of the researchers and their crystallographic background. Something very similar to the quasicrystalline structure, indeed, had been already discovered few decades ago and not recognized. I am referring to natural periodic approximants, crystalline solids with similar chemical composition to a quasicrystal, but whose atomic arrangement is slightly distorted so that the symmetry conforms to the conventional laws of three-dimensional crystallography. For example, the mineral naquite, FeSi, was found and described from

© The Author(s), under exclusive license to Springer Nature Switzerland AG 2020
L. Bindi, *Natural Quasicrystals*, SpringerBriefs in Crystallography,
https://doi.org/10.1007/978-3-030-45677-1_9

Fig. 9.1 Idealized structure
of the simplest cubic
approximant. This represents
the naquite and brownleeite
structure

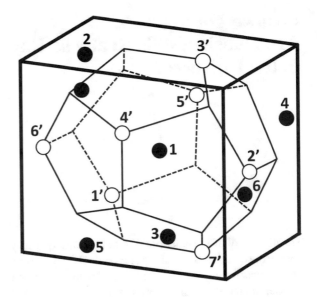

a lunar meteorite (Anand et al. 2004), and the mineral brownleeite, MnSi, was found
in cometary interplanetary dust particles (Nakamura-Messenger et al. 2010). Both
these minerals crystallize with the structure of the simplest, idealized known cubic
approximant: $a \approx 4.6$ Å; space-group type $P2_13$; 8 atoms per unit cell in the $4a$
Wyckoff positions.

The M (Fe or Mn) and Si atoms of such an idealized structure have the coordinates
(x,x,x) with $x_M = (\tau - 1)/4 \approx 0.155$ and $x_{Si} = 1 - x \approx 0.845$. As a result, each M atom
is surrounded by seven Si atoms at a subset of vertices of a dodecahedron, and vice
versa (Fig. 9.1). Looking more carefully, we can see that parts of MacKay icosahedra
can also be observed (one unit-cell is too small to house a whole icosahedron). Such
a simple $P2_13$ MSi structure produces a system of Bragg reflections similar to that
observed in true icosahedral quasicrystals. According to Dmitrienko (1990), the MSi
structure contains both parts of higher- and low-order approximants; the approximant
of lower order presents a standard body-centered structure, and therefore in MSi one
can find 8-atom clusters of the body-centered structure with atoms at 000, 001, 100,
010, ½, ½, ½, −½, ½, ½, ½, ½, −½, and ½,−½,½.

Noteworthy, the MSi structure can be obtained by projecting the crystallographic
orbit of a six-dimensional cubic structure with one atom per unit cell. If the strip
in the six-dimensional space is selected in a suitable way, the resulting structure
will be periodic; other positions of the strip generate higher-order approximants and
quasicrystals.

So, as a final remark, we do not know for certain whether quasicrystals are rare
or common in the universe, but the discovery of natural quasicrystals forces us to
set aside the historic arguments that suggested they must be rare. Scientists will
learn more as they conduct further searches for natural quasicrystals and perform the
experiments they inspire.

References

Anand M, Taylor LA, Nazarov MA, Shu J, Mao H-K, Hemley RJ (2004) Space weathering on airless planetary bodies: clues from the lunar mineral hapkeite. Proc Nat Acad Sci USA 101:6847–6851

Belyanin GA, Kramers JD, Andreoli MAG, Greco F, Gucsik A, Makhubela TV, Przybylowicz WJ, Wiedenbeck M (2018) Petrography of the carbonaceous, diamond-bearing stone "Hypatia" from southwest Egypt: a contribution to the debate on its origin. Geoch Cosmoch Acta 223:462–492

Bindi L, Steinhardt PJ, Yao N, Lu PJ (2009) Natural quasicrystals. Science 324:1306–1309

Dmitrienko VE (1990) Cubic approximants in quasicrystals structure. J Phys France 51:2717–2732

Gornostaeva TA, Mokhov AV, Kartashov PM, Bogatikov OA (2018) Impactor type and model of the origin of the zhamanshin astrobleme, Kazakhstan. Petrology 26:82–95

Nakamura-Messenger K, Keller L, Clemett SJ, Messenger S, Jones JH, Palma RL, Pepin RO, Klock W, Zolensky ME, Tatsuoka H (2010) Brownleeite: a new manganese silicide mineral in an interplanetary dust particle. Am Mineral 95:221–228

Suttle M, Twegar K, Nava J, Spiess R, Spratt J, Campanale F, Folco L (2019) A unique CO-like micrometeorite hosting an exotic Al-Cu-Fe-bearing assemblage—close affinities with the Khatyrka meteorite. Sci Rep 9:12426

Xie X, Chen M (2016) Shock-induced redistribution of trace elements. In Suizhou meteorite: mineralogy and shock metamorphism. Springer, Berlin, pp 211–222

Printed in the United States
By Bookmasters